Vector Bundles and Their Applications

T0191691

Mathematics and Its Applications

Managing Editor:

M. HAZEWINKEL
Centre for Mathematics and Computer Science, Amsterdam, The Netherlands

Volume 447

Vector Bundles and Their Applications

by

Glenys Luke

St. Hugh's College,
Oxford University,
Oxford, U.K.

and

Alexander S. Mishchenko

Department of Mathematics,
Moscow State University,
Moscow, Russia

KLUWER ACADEMIC PUBLISHERS
DORDRECHT / BOSTON / LONDON

A C.I.P. Catalogue record for this book is available from the Library of Congress.

ISBN 978-1-4419-4802-1

Published by Kluwer Academic Publishers,
P.O. Box 17, 3300 AA Dordrecht, The Netherlands.

Sold and distributed in North, Central and South America
by Kluwer Academic Publishers,
101 Philip Drive, Norwell, MA 02061, U.S.A.

In all other countries, sold and distributed
by Kluwer Academic Publishers,
P.O. Box 322, 3300 AH Dordrecht, The Netherlands.

Printed on acid-free paper

CONTENTS

v

PREFACE

In the last few years the use of geometric methods has permeated many more branches of mathematics and the sciences. Briefly its role may be characterized as follows. Whereas methods of mathematical analysis describe phenomena 'in the small', geometric methods contribute to giving the picture 'in the large'. A second no less important property of geometric methods is the convenience of using its language to describe and give qualitative explanations for diverse mathematical phenomena and patterns. From this point of view, the theory of vector bundles together with mathematical analysis on manifolds (global analysis and differential geometry) has provided a major stimulus. Its language turned out to be extremely fruitful: connections on principal vector bundles (in terms of which various field theories are described), transformation groups including the various symmetry groups that arise in connection with physical problems, in asymptotic methods of partial differential equations with small parameter, in elliptic operator theory, in mathematical methods of classical mechanics and in mathematical methods in economics. There are other currently less significant applications in other fields. Over a similar period, university education has changed considerably with the appearance of new courses on differential geometry and topology. New textbooks have been published but 'geometry and topology' has not, in our opinion, been well covered from a practical applications point of view. Existing monographs on vector bundles have been mainly of a purely theoretical nature, devoted to the internal geometric and topological problems of the subject. Students from related disciplines have found the texts difficult to use. It therefore seems expedient to have a simpler book containing numerous illustrations and applications to various problems in mathematics and the sciences.

Part of this book is based on material contained in lectures of the author, A.Mishchenko, given to students of the Mathematics Department at Moscow

State University and is a revised version of the Russian edition of 1984. Some of the less important theorems have been omitted and some proofs simplified and clarified. The focus of attention was towards explaining the most important notions and geometric constructions connected with the theory of vector bundles.

Theorems were not always formulated in maximal generality but rather in such a way that the geometric nature of the objects came to the fore. Whenever possible examples were given to illustrate the role of vector bundles. Thus the book contains sections on locally trivial bundles, and on the simplest properties and operations on vector bundles. Further properties of a homotopic nature, including characteristic classes, are also expounded. Considerable attention is devoted to natural geometric constructions and various ways of constructing vector bundles. Basic algebraic notions involved in describing and calculating K-theory are studied and the particularly interesting field of applications to the theory of elliptic pseudodifferential operators is included. The exposition finishes with further applications of vector bundles to topology. Certain aspects which are well covered in other sources have been omitted in order to prevent the book becoming too bulky.

1

INTRODUCTION TO THE
LOCALLY TRIVIAL BUNDLES
THEORY

1.1 LOCALLY TRIVIAL BUNDLES

The definition of a locally trivial bundle was coined to capture an idea which recurs in a number of different geometric situations. We commence by giving a number of examples.

The surface of the cylinder can be seen as a disjoint union of a family of line segments continuously parametrized by points of a circle. The Möbius band can be presented in similar way. The two dimensional torus embedded in the three dimensional space can presented as a union of a family of circles (meridians) parametrized by points of another circle (a parallel).

Now, let M be a smooth manifold embedded in the Euclidean space $\mathbf{R}^N$ and TM the space embedded in $\mathbf{R}^N \times \mathbf{R}^N$, the points of which are the tangent vectors of the manifold M. This new space TM can be also be presented as a union of subspaces $T_x M$, where each $T_x M$ consists of all the tangent vectors to the manifold M at the single point x. The point x of M can be considered as a parameter which parametrizes the family of subspaces $T_x M$. In all these cases the space may be partitioned into fibers parametrized by points of the base.

The examples considered above share two important properties: a) any two fibers are homeomorphic, b) despite the fact that the whole space cannot be presented as a Cartesian product of a fiber with the base (the parameter space), if we restrict our consideration to some small region of the base the part of the fiber space over this region is such a Cartesian product. The two properties above are the basis of the following definition.

1

Definition 1 *Let E and B be two topological spaces with a continuous map*

$$p : E \longrightarrow B.$$

The map p is said to define a locally trivial bundle if there is a topological space F such that for any point $x \in B$ there is a neighborhood $U \ni x$ for which the inverse image $p^{-1}(U)$ is homeomorphic to the Cartesian product $U \times F$. Moreover, it is required that the homeomorphism

$$\varphi : U \times F \longrightarrow p^{-1}(U)$$

preserves fibers, it is a 'fiberwise' map, that is, the following equality holds:

$$p(\varphi(x, f)) = x, x \in U, f \in F.$$

The space E is called total space of the bundle or the fiberspace , the space B is called the base of the bundle , the space F is called the fiber of the bundle and the mapping p is called the projection . The requirement that the homeomorphism φ be fiberwise means in algebraic terms that the diagram

$$
\begin{array}{ccc}
U \times F & \xrightarrow{\varphi} & p^{-1}(U) \\
\downarrow{\scriptstyle \pi} & & \downarrow{\scriptstyle p} \\
U & = & U
\end{array}
$$

where

$$\pi : U \times F \longrightarrow U, \quad \pi(x, f) = x$$

is the projection onto the first factor is commutative.

One problem in the theory of fiber spaces is to classify the family of all locally trivial bundles with fixed base B and fiber F. Two locally trivial bundles $p : E \longrightarrow B$ and $p' : E' \longrightarrow B$ are considered to be isomorphic if there is a homeomorphism $\psi : E \longrightarrow E'$ such that the diagram

$$
\begin{array}{ccc}
E & \xrightarrow{\psi} & E' \\
\downarrow{\scriptstyle p} & & \downarrow{\scriptstyle p'} \\
B & = & B
\end{array}
$$

is commutative. It is clear that the homeomorphism ψ gives a homeomorphism of fibers $F \longrightarrow F'$. To specify a locally trivial bundle it is not necessary to be given the total space E explicitly. It is sufficient to have a base B, a fiber F and a family of mappings such that the total space E is determined 'uniquely' (up to isomorphisms of bundles). Then according to the definition of a locally trivial bundle, the base B can be covered by a family of open sets $\{U_\alpha\}$ such that each inverse image $p^{-1}(U_\alpha)$ is fiberwise homeomorphic to $U_\alpha \times F$. This gives a system of homeomorphisms

$$\varphi_\alpha : U_\alpha \times F \longrightarrow p^{-1}(U_\alpha).$$

Since the homeomorphisms φ_α preserve fibers it is clear that for any open subset $V \subset U_\alpha$ the restriction of φ_α to $V \times F$ establishes the fiberwise homeomorphism of $V \times F$ onto $p^{-1}(V)$. Hence on $U_\alpha \times U_\beta$ there are two fiberwise homeomorphisms

$$\varphi_\alpha : (U_\alpha \cap U_\beta) \times F \longrightarrow p^{-1}(U_\alpha \cap U_\beta),$$
$$\varphi_\beta : (U_\alpha \cap U_\beta) \times F \longrightarrow p^{-1}(U_\alpha \cap U_\beta).$$

Let $\varphi_{\alpha\beta}$ denote the homeomorphism $\varphi_\beta^{-1}\varphi_\alpha$ which maps $(U_\alpha \cap U_\beta) \times F$ onto itself. The locally trivial bundle is uniquely determined by the following collection: the base B, the fiber F, the covering U_α and the homeomorphisms

$$\varphi_{\alpha\beta} : (U_\alpha \cap U_\beta) \times F \longrightarrow (U_\alpha \cap U_\beta) \times F.$$

The total space E should be thought of as a union of the Cartesian products $U_\alpha \times F$ with some identifications induced by the homeomorphisms $\varphi_{\alpha\beta}$. By analogy with the terminology for smooth manifolds, the open sets U_α are called *charts* , the family $\{U_\alpha\}$ is called *the atlas of charts* , the homeomorphisms φ_α are called *the coordinate homeomorphisms* and the $\varphi_{\alpha\beta}$ are called *the transition functions* or *the sewing functions* . Sometimes the collection $\{U_\alpha, \varphi_\alpha\}$ is called the atlas. Thus any atlas determines a locally trivial bundle. Different atlases may define isomorphic bundles but, beware, not any collection of homeomorphisms φ_α forms an atlas. For the classification of locally trivial bundles, families of homeomorphisms $\varphi_{\alpha\beta}$ that actually determine bundles should be selected and then separated into classes which determine isomorphic bundles. For the homeomorphisms $\varphi_{\alpha\beta}$ to be transition functions for some locally trivial bundle:

$$\varphi_{\beta\alpha} = \varphi_\beta^{-1}\varphi_\alpha. \tag{1.1}$$

Then for any three indices α, β, γ on the intersection $(U_\alpha \cap U_\beta \cap U_\gamma) \times F$ the following relation holds:

$$\varphi_{\alpha\gamma}\varphi_{\gamma\beta}\varphi_{\beta\alpha} = \mathbf{Id},$$

where $\mathbf{Id}$ is the identity homeomorphism and for each α,

$$\varphi_{\alpha\alpha} = \mathbf{Id}. \tag{1.2}$$

In particular
$$\varphi_{\alpha\beta}\varphi_{\beta\alpha} = \mathbf{Id},\qquad(1.3)$$
thus
$$\varphi_{\alpha\beta} = \varphi_{\beta\alpha}^{-1}.$$
Hence for an atlas the $\varphi_{\alpha\beta}$ should satisfy
$$\varphi_{\alpha\alpha} = \mathbf{Id},\quad \varphi_{\alpha\gamma}\varphi_{\gamma\beta}\varphi_{\beta\alpha} = \mathbf{Id}.\qquad(1.4)$$
These conditions are sufficient for a locally trivial bundle to be reconstructed from the base B, fiber F, atlas $\{U_\alpha\}$ and homeomorphisms $\varphi_{\beta\alpha}$. To see this, let
$$E' = \cup_\alpha(U_\alpha \times F)$$
be the disjoint union of the spaces $U_\alpha \times F$. Introduce the following equivalence relation: the point $(x, f) \in U_\alpha \times F$ is related to the point $(y, g) \in U_\beta \times F$ iff
$$x = y \in U_\alpha \cap U_\beta$$
and
$$(y, g) = \varphi_{\beta\alpha}(x, f).$$
The conditions (1.2), (1.3) guarantee that this is an equivalence relation, that is, the space E' is partitioned into disjoint classes of equivalent points. Let E be the quotient space determined by this equivalence relation, that is, the set whose points are equivalence classes. Give E the quotient topology with respect to the projection
$$\pi : E' \longrightarrow E$$
which associates to a (x, f) its the equivalence class. In other words, the subset $G \subset E$ is called open iff $\pi^{-1}(G)$ is open set. There is the natural mapping p' from E' to B:
$$p'(x, f) = x.$$
Clearly the mapping p' is continuous and equivalent points maps to the same image. Hence the mapping p' induces a map
$$p : E \longrightarrow B$$
which associates to an equivalence class the point assigned to it by p'. The mapping p is continuous. It remains to construct the coordinate homeomorphisms. Put
$$\varphi_\alpha = \pi|_{U_\alpha \times F} : U_\alpha \times F \longrightarrow E.$$
Each class $z \in p^{-1}(U_\alpha)$ has a unique representative $(x, f) \in U_\alpha \times F$. Hence φ_α is a one to one mapping onto $p^{-1}(U_\alpha)$. By virtue of the quotient topology

on E the mapping φ_α is a homeomorphism. It is easy to check that (compare with (1.1))

$$\varphi_\beta^{-1}\varphi_\alpha = \varphi_{\beta\alpha}.$$

So we have shown that locally trivial bundles may be defined by atlas of charts $\{U_\alpha\}$ and a family of homeomorphisms $\{\varphi_{\beta\alpha}\}$ satisfying the conditions (1.2), (1.3). Let us now determine when two atlases define isomorphic bundles. First of all notice that if two bundles $p : E \longrightarrow B$ and $p' : E' \longrightarrow B$ with the same fiber F have the same transition functions $\{\varphi_{\beta\alpha}\}$ then these two bundles are isomorphic. Indeed, let

$$\varphi_\alpha : U_\alpha \longrightarrow p^{-1}(U_\alpha).$$

$$\psi_\alpha : U_\alpha \longrightarrow p'^{-1}(U_\alpha).$$

be the corresponding coordinate homeomorphisms and assume that

$$\varphi_{\beta\alpha} = \varphi_\beta^{-1}\varphi_\alpha = \psi_\beta^{-1}\psi_\alpha = \psi_{\beta\alpha}.$$

Then

$$\varphi_\alpha\psi_\alpha^{-1} = \varphi_\beta\psi_\beta^{-1}.$$

We construct a homeomorphism

$$\psi : E' \longrightarrow E.$$

Let $x \in E'$. The atlas $\{U_\alpha\}$ covers the base B and hence there is an index α such that $x \in p'^{-1}(U_\alpha)$. Set

$$\psi(x) = \varphi_\alpha\psi_\alpha^{-1}(x).$$

It is necessary to establish that the value of $\psi(x)$ is independent of the choice of index α. If $x \in p'^{-1}(U_\beta)$ also then

$$\varphi_\beta\psi_\beta^{-1}(x) = \varphi_\alpha\varphi_\alpha^{-1}\varphi_\beta\psi_\beta^{-1}\psi_\alpha\psi_\alpha^{-1}(x) =$$
$$= \varphi_\alpha\varphi_{\alpha\beta}\varphi_{\beta\alpha}\psi_\alpha^{-1}(x) = \varphi_\alpha\psi_\alpha^{-1}(x).$$

Hence the definition of $\psi(x)$ is independent of the choice of chart. Continuity and other necessary properties are evident. Further, given an atlas $\{U_\alpha\}$ and coordinate homomorphisms $\{\varphi_\alpha\}$, if $\{V_\beta\}$ is a finer atlas (that is, $V_\beta \subset U_\alpha$ for some $\alpha = \alpha(\beta)$) then for the atlas $\{V_\beta\}$, the coordinate homomorphisms are defined in a natural way

$$\varphi_\beta' = \varphi_{\alpha(\beta)}|_{(V_\beta \times F)} : V_\beta \times F \longrightarrow p^{-1}(V_\beta).$$

The transition functions $\varphi_{\beta_1,\beta_2}'$ for the new atlas $\{V_\beta\}$ are defined using restrictions

$$\varphi_{\beta_1,\beta_2}' = \varphi_{\alpha(\beta_1),\alpha(\beta_2)}|_{(V_{\beta_1} \cap V_{\beta_2}) \times F} : (V_{\beta_1} \cap V_{\beta_2}) \times F \longrightarrow (V_{\beta_1} \cap V_{\beta_2}) \times F.$$

Thus if there are two atlases and transition functions for two bundles, with a common refinement, that is, a finer atlas with transition functions given by restrictions, it can be assumed that the two bundles have the same atlas. If $\varphi_{\beta\alpha}$, $\varphi'_{\beta\alpha}$ are two systems of the transition functions (for the same atlas), giving isomorphic bundles then the transition functions $\varphi_{\beta\alpha}$, $\varphi'_{\beta\alpha}$ must be related.

Theorem 1 *Two systems of the transition functions $\varphi_{\beta\alpha}$, and $\varphi'_{\beta\alpha}$ define isomorphic locally trivial bundles iff there exist fiber preserving homeomorphisms*

$$h_\alpha : U_\alpha \times F \longrightarrow U_\alpha \times F$$

such that

$$\varphi_{\beta\alpha} = h_\beta^{-1} \varphi'_{\beta\alpha} h_\alpha. \tag{1.5}$$

Proof.

Suppose that two bundles $p : E \longrightarrow B$ and $p' : E' \longrightarrow B$ with the coordinate homeomorphisms φ_α and φ'_α are isomorphic. Then there is a homeomorphism $\psi : E' \longrightarrow E$. Let

$$h_\alpha = \varphi'^{-1}_\alpha \psi^{-1} \varphi_\alpha.$$

Then

$$h_\beta^{-1} \varphi'_{\beta\alpha} h_\alpha = \varphi_\beta^{-1} \psi \varphi'_\beta \varphi'_{\beta\alpha} \varphi'^{-1}_\alpha \psi^{-1} \varphi_\alpha =$$
$$= \varphi_\beta^{-1} \psi \varphi'_\beta \varphi'^{-1}_\beta \varphi'_\alpha \varphi'^{-1}_\alpha \psi^{-1} \varphi_\alpha = \varphi_{\beta\alpha}.$$

Conversely, if the relation (1.5) holds, put

$$\psi = \varphi_\alpha h_\alpha^{-1} \varphi'^{-1}_\alpha. \tag{1.6}$$

The definition (1.6) is valid for the subspaces $p'^{-1}(U_\alpha)$ covering E'. To prove that the right hand sides of (1.6) coincide on the intersection $p'^{-1}(U_\alpha \cap U_\beta)$ the relations (1.5) are used:

$$\varphi_\beta h_\beta^{-1} \varphi'^{-1}_\beta = \varphi_\alpha \varphi_\alpha^{-1} \varphi_\beta h_\beta^{-1} \varphi'^{-1}_\beta \varphi'_\alpha \varphi'^{-1}_\alpha =$$
$$= \varphi_\alpha \varphi_{\alpha\beta}^{-1} h_\beta^{-1} \varphi'^{-1}_{\beta\alpha} \varphi'^{-1}_\alpha = \varphi_\alpha h_\alpha^{-1} \varphi'^{-1}_{\alpha\beta} h_\beta h_\beta^{-1} \varphi'^{-1}_{\beta\alpha} \varphi'^{-1}_\alpha =$$
$$= \varphi_\alpha h_\alpha^{-1} \varphi'^{-1}_\alpha.$$

■

Examples

1. Let $E = B \times F$ and $p : E \longrightarrow B$ be projections onto the first factors. Then the atlas consists of one chart $U_\alpha = B$ and only one the transition function $\varphi_{\alpha\alpha} = \mathbf{Id}$ and the bundle is said to be *trivial* .

2. Let E be the Möbius band. One can think of this bundle as a square in the plane, $\{(x, y) : 0 \leq x \leq 1, 0 \leq y \leq 1\}$ with the points $(0, y)$ and $(1, 1 - y)$ identified for each $y \in [0, 1]$. The projection p maps the space E onto the segment $I_x = \{0 \leq x \leq 1\}$ with the endpoints $x = 0$ and $x = 1$ identified, that is, onto the circle S^1. Let us show that the map p defines a locally trivial bundle. The atlas consists of two intervals (recall 0 and 1 are identified)

$$U_\alpha = \{0 < x < 1\}, \ U_\beta = \{0 \leq x < \frac{1}{2}\} \cup \{\frac{1}{2} < x \leq 1\}.$$

The coordinate homeomorphisms may be defined as following:

$$\varphi_\alpha : U_\alpha \times I_y \longrightarrow E, \varphi_\alpha(x, y) = (x, y),$$
$$\varphi_\beta : U_\beta \times I_y \longrightarrow E$$
$$\varphi_\beta(x, y) = (x, y) \quad \text{for} \quad 0 \leq x < \frac{1}{2},$$
$$\varphi_\beta(x, y) = (x, 1 - y) \quad \text{for} \quad \frac{1}{2} < x \leq 1.$$

The intersection of two charts $U_\alpha \cap U_\beta$ consists of union of two intervals $U_\alpha \cap U_\beta = (0, \frac{1}{2}) \cup (\frac{1}{2}, 1)$. The transition function $\varphi_{\beta\alpha}$ have the following form

$$\varphi_{\beta\alpha} = (x, y) \text{ for } 0 < x < \frac{1}{2},$$
$$\varphi_{\beta\alpha} = (x, 1 - y) \text{ for } \frac{1}{2} < x < 1.$$

The Möbius band is not isomorphic to a trivial bundle. Indeed, for a trivial bundle all transition functions can be chosen equal to the identity. Then by Theorem 1 there exist fiber preserving homeomorphisms

$$h_\alpha : U_\alpha \times I_y \longrightarrow U_\alpha \times I_y,$$
$$h_\beta : U_\beta \times I_y \longrightarrow U_\beta \times I_y,$$

such that

$$\varphi_{\beta\alpha} = h_\beta^{-1} h_\alpha$$

in its domain of definition $(U_\alpha \cap U_\beta) \times I_y$. Then h_α, h_β are fiberwise homeomorphisms for fixed value of the first argument x giving homeomorphisms of interval I_y to itself. Each homeomorphism of the interval to itself maps end points to end points. So the functions

$$h_\alpha(x,0), \ h_\alpha(x,1), \ h_\beta(x,0), \ h_\beta(x,1)$$

are constant functions, with values equal to zero or one. The same is true for the functions $h_\beta^{-1}h_\alpha(x,0)$. On the other hand the function $\varphi_{\beta\alpha}(x,0)$ is not constant because it equals zero for each $0 < x < \frac{1}{2}$ and equals one for each $\frac{1}{2} < x < 1$. This contradiction shows that the Möbius band is not isomorphic to a trivial bundle.

3. Let E be the space of tangent vectors to two dimensional sphere $\mathbf{S}^2$ embedded in three dimensional Euclidean space $\mathbf{R}^3$. Let

$$p : E \longrightarrow \mathbf{S}^2$$

be the map associating each vector to its initial point. Let us show that p is a locally trivial bundle with fiber $\mathbf{R}^2$. Fix a point $s_0 \in \mathbf{S}^2$. Choose a Cartesian system of coordinates in $\mathbf{R}^3$ such that the point s_0 is the North Pole on the sphere (that is, the coordinates of s_0 equal $(0,0,1)$). Let U be the open subset of the sphere $\mathbf{S}^2$ defined by inequality $z > 0$. If $s \in U$, $s = (x, y, z)$, then

$$x^2 + y^2 + z^2 = 1, \ z > 0.$$

Let $\vec{e} = (\xi, \eta, \zeta)$ be a tangent vector to the sphere at the point s. Then

$$x\xi + y\eta + z\zeta = 0,$$

that is,

$$\zeta = -(x\xi + y\eta)/z.$$

Define the map

$$\varphi : U \times \mathbf{R}^2 \longrightarrow p^{-1}(U)$$

by the formula

$$\varphi(x, y, z, \xi, \eta) = (x, y, z, \xi, \eta, -(x\xi + y\eta)/z)$$

giving the coordinate homomorphism for the chart U containing the point $s_0 \in \mathbf{S}^2$. Thus the map p gives a locally trivial bundle. This bundle is called the *tangent bundle* of the sphere $\mathbf{S}^2$.

1.2 THE STRUCTURE GROUPS OF THE LOCALLY TRIVIAL BUNDLES

The relations (1.4,1.5) obtained in the previous section for the transition functions of a locally trivial bundle are similar to those involved in the calculation of one dimensional cohomology with coefficients in some algebraic sheaf. This analogy can be explain after a slight change of terminology and notation and the change will be useful for us for investigating the classification problem of locally trivial bundles. Notice that a fiberwise homeomorphism of the Cartesian product of the base U and the fiber F onto itself

$$\varphi : U \times F \longrightarrow U \times F, \qquad (1.7)$$

can be represented as a family of homeomorphisms of the fiber F onto itself, parametrized by points of the base B. In other words, each fiberwise homeomorphism φ defines a map

$$\bar\varphi : U \longrightarrow \textbf{Homeo (F)}, \qquad (1.8)$$

where **Homeo (F)** is the group of all homeomorphisms of the fiber F. Furthermore, if we choose the right topology on the group **Homeo (F)** the map $\bar\varphi$ becomes continuous. Sometimes the opposite is true: the map (1.8) generates the fiberwise homeomorphism (1.7) with respect to the formula

$$\varphi(x, f) = (x, \bar\varphi(x)f).$$

So instead of $\varphi_{\alpha\beta}$ a family of functions

$$\bar\varphi_{\alpha\beta} : U_\alpha \cap U_\beta \longrightarrow \textbf{Homeo (F)},$$

can be defined on the intersection $U_\alpha \cap U_\beta$ and having values in the group **Homeo (F)**. In homological algebra the family of functions $\bar\varphi_{\alpha\beta}$ is called a *one dimensional cochain* with values in the sheaf of germs of functions with values in the group **Homeo (F)**. The condition (1.4) from the section 1.1 means that

$$\bar\varphi_{\alpha\alpha}(x) = \textbf{Id},$$
$$\bar\varphi_{\alpha\gamma}(x)\bar\varphi_{\gamma\beta}(x)\bar\varphi_{\beta\alpha}(x) = \textbf{Id}.$$
$$x \in U_\alpha \cap U_\beta \cap U_\gamma.$$

and we say that the cochain $\{\bar\varphi_{\alpha\beta}\}$ is a *cocycle*. The condition (1.5) means that there is a zero dimensional cochain $h_\alpha : U_\alpha \longrightarrow \textbf{Homeo (F)}$ such that

$$\bar\varphi_{\beta\alpha}(x) = h_\beta^{-1}(x)\bar\varphi'_{\beta\alpha}(x)h_\alpha(x), \; x \in U_\alpha \cap U_\beta.$$

Using the language of homological algebra the condition (1.5) means that co-cycles $\{\bar{\varphi}_{\beta\alpha}\}$ and $\{\bar{\varphi}'_{\beta\alpha}\}$ are cohomologous. Thus the family of locally trivial bundles with fiber F and base B is in one to one correspondence with the one dimensional cohomology of the space B with coefficients in the sheaf of the germs of continuous **Homeo** (**F**)–valued functions for given open cover-ing $\{U_\alpha\}$. Despite obtaining a simple description of the family of locally trivial bundles in terms of homological algebra, it is ineffective since there is no simple method of calculating cohomologies of this kind. Nevertheless, this representa-tion of the transition functions as a cocycle turns out very useful because of the situation described below. First of all notice that using the new interpretation a locally trivial bundle is determined by the base B, the atlas $\{U_\alpha\}$ and the functions $\{\varphi_{\alpha\beta}\}$ taking the value in the group $G = $ **Homeo** (**F**). The fiber F itself does not directly take part in the description of the bundle. Hence, one can at first describe a locally trivial bundle as a family of functions $\{\varphi_{\alpha\beta}\}$ with values in some topological group G, and after that construct the total space of the bundle with fiber F by additionally defining an action of the group G on the space F, that is, defining a continuous homomorphism of the group G into the group **Homeo** (**F**). Secondly, the notion of locally trivial bundle can be generalized and the structure of bundle made richer by requiring that both the transition functions $\bar{\varphi}_{\alpha\beta}$ and the functions h_α are not arbitrary but take values in some subgroup of the homeomorphism group **Homeo** (**F**).Thirdly, some-times information about locally trivial bundle may be obtained by substituting some other fiber F' for the fiber F but using the 'same' transition functions. Thus we come to a new definition of a locally trivial bundle with additional structure — the group where the transition functions take their values.

Definition 2 *Let E, B, F be topological spaces and G be a topological group which acts continuously on the space F. A continuous map*

$$p : E \longrightarrow B$$

is said to be a locally trivial bundle with fiber F and the structure group G if there is an atlas $\{U_\alpha\}$ and the coordinate homeomorphisms

$$\varphi_\alpha : U_\alpha \times F \longrightarrow p^{-1}(U_\alpha)$$

such that the transition functions

$$\varphi_{\beta\alpha} = \varphi_\beta^{-1}\varphi_\alpha : (U_\alpha \cap U_\beta) \times F \longrightarrow (U_\alpha \cap U_\beta) \times F$$

have the form

$$\varphi_{\beta\alpha}(x, f) = (x, \bar{\varphi}_{\beta\alpha}(x)f),$$

where $\bar{\varphi}_{\beta\alpha} : (U_\alpha \cap U_\beta) \longrightarrow G$ are continuous functions satisfying the conditions

$$\bar{\varphi}_{\alpha\alpha}(x) \equiv 1, \ x \in U_\alpha,$$
$$\bar{\varphi}_{\alpha\beta}(x)\bar{\varphi}_{\beta\gamma}(x)\bar{\varphi}_{\gamma\alpha}(x) \equiv 1, \ x \in U_\alpha \cap U_\beta \cap U_\gamma. \tag{1.9}$$

The functions $\bar{\varphi}_{\alpha\beta}$ are also called the transition functions

Let

$$\psi : E' \longrightarrow E$$

be an isomorphism of locally trivial bundles with the structure group G. Let φ_α and φ'_α be the coordinate homeomorphisms of the bundles $p : E \longrightarrow B$ and $p' : E' \longrightarrow B$, respectively. One says that the isomorphism ψ is *compatible with the structure group G* if the homomorphisms

$$\varphi_\alpha^{-1}\psi\varphi'_\alpha : U_\alpha \times F \longrightarrow U_\alpha \times F$$

are determined by continuous functions

$$h_\alpha : U_\alpha \longrightarrow G,$$

defined by relation

$$\varphi_\alpha^{-1}\psi\varphi'_\alpha(x, f) = (x, h_\alpha(x)f). \tag{1.10}$$

Thus two bundles with the structure group G and transition functions $\bar{\varphi}_{\beta\alpha}$ and $\varphi'_{\beta\alpha}$ are isomorphic, the isomorphism being compatible with the structure group G, if

$$\bar{\varphi}_{\beta\alpha}(x) = h_\beta(x)\bar{\varphi}'_{\beta\alpha}(x)h_\alpha(x) \tag{1.11}$$

for some continuous functions $h_\alpha : U_\alpha \longrightarrow G$. So two bundles whose the transition functions satisfy the condition (1.11) are called *equivalent bundles* . It is sometimes useful to increase or decrease the structure group G. Two bundles which are not equivalent with respect of the structure group G may become equivalent with respect to a larger structure group G', $G \subset G'$. When a bundle with the structure group G admits transition functions with values in a subgroup H, it is said that the structure group G is *reduced to subgroup H*. It is clear that if the structure group of the bundle $p : E \longrightarrow B$ consists of only one element then the bundle is trivial. So to prove that the bundle is trivial, it is sufficient to show that its the structure group G may be reduced to the trivial subgroup. More generally, if

$$\rho : G \longrightarrow G'$$

is a continuous homomorphism of topological groups and we are given a locally trivial bundle with the structure group G and the transition functions

$$\varphi_{\alpha\beta} : U_\alpha \cap U_\beta \longrightarrow G$$

then a new locally trivial bundle may be constructed with structure group G' for which the transition functions are defined by

$$\varphi'_{\alpha\beta}(x) = \rho(\varphi_{\alpha\beta}(x)).$$

This operation is called *a change of the structure group* (with respect to the homomorphism ρ).

Remark

Note that the fiberwise homeomorphism

$$\varphi : U \times F \longrightarrow U \times F$$

in general is not induced by continuous map

$$\bar{\varphi} : U \longrightarrow \textbf{Homeo (F)}. \qquad (1.12)$$

Because of lack of space we will not analyze the problem and note only that later on in all our applications the fiberwise homeomorphisms will be induced by the continuous maps (1.12) into the structure group G.

Now we can return to the third situation, that is, to the possibility to choosing a space as a fiber of a locally trivial bundle with the structure group G. Let us consider the fiber

$$F = G$$

with the action of G on F being that of left translation, that is, the element $g \in G$ acts on the F by the homeomorphism

$$g(f) = gf, \ f \in F = G.$$

Definition 3 *A locally trivial bundle with the structure group G is called principal G–bundle if $F = G$ and action of the group G on F is defined by the left translations.*

An important property of principal G–bundles is the consistency of the homeomorphisms with the structure group G and it can be described not only in terms of the transition functions (the choice of which is not unique) but also in terms of equivariant properties of bundles.

Theorem 2 *Let*
$$p : E \longrightarrow B$$
be a principal G–bundle,
$$\varphi_\alpha : U_\alpha \times G \longrightarrow p^{-1}(U_\alpha)$$
be the coordinate homeomorphisms. Then there is a right action of the group G on the total space E such that:

1. *the right action of the group G is fiberwise, that is,*

$$p(x) = p(xg), \ x \in E, \ g \in G.$$

2. *the homeomorphism φ_α^{-1} transforms the right action of the group G on the total space into right translations on the second factor, that is,*

$$\varphi_\alpha(x, f)g = f_\alpha(x, fg), \ x \in U_\alpha, \ f, g \in G. \tag{1.13}$$

Proof.

According to the definitions 2 and 3, the transition functions $\varphi_{\beta\alpha} = \varphi_\beta^{-1} \varphi_\alpha$ have the following form

$$\varphi_{\beta\alpha}(x, f) = (x, \bar{\varphi}_{\beta\alpha}(x)f),$$

where
$$\bar{\varphi}_{\beta\alpha} : U_\alpha \cap U_\beta \longrightarrow G$$

are continuous functions satisfying the conditions (1.9). Since an arbitrary point $z \in E$ can be represented in the form

$$z = \varphi_\alpha(x, f)$$

for some index α, the formula (1.13) determines the continuous right action of the group G provided that this definition is independent of the choice of index α. So suppose that
$$z = \varphi_\alpha(x, f) = \varphi_\beta(x, f').$$

We need to show that the element zg does not depend on the choice of index, that is,
$$\varphi_\alpha(x, fg) = \varphi_\beta(x, f'g)$$

or
$$(x, f'g) = \varphi_{\alpha^{-1}}\varphi_\beta(x, fg) = \varphi_{\beta\alpha}(x, fg)$$
or
$$f'g = \bar\varphi_{\beta\alpha}(x)fg. \tag{1.14}$$
However,
$$(x, f') = \varphi_\beta^{-1}\varphi_\alpha(x, f) = \varphi_{\beta\alpha}(x, f) = (x, \bar\varphi_{\beta\alpha}f),$$
Hence
$$f' = \bar\varphi_{\beta\alpha}(x)f. \tag{1.15}$$
Thus multiplying (1.15) by g on the right gives (1.14).   ∎

Theorem 2 allows us to consider principal G–bundles as having a right action on the total space.

Theorem 3 *Let*
$$\psi : E' \longrightarrow E \tag{1.16}$$
be a fiberwise map of principal G-bundles. The map (1.16) is the isomorphism of locally trivial bundles with the structure group G, that is, compatible with the structure group G iff this map is equivariant (with respect to right actions of the group G on the total spaces).

Proof.

Let
$$p : E \longrightarrow B,$$
$$p' : E' \longrightarrow B$$
be locally trivial principal bundles both with the structure group G and let φ_α, φ'_α be coordinate homeomorphisms. Then by the definition (1.10), the map ψ is an isomorphism of locally trivial bundles with structure group G when
$$\varphi_\alpha^{-1}\psi\varphi'_\alpha(x, g) = (x, h_\alpha(x)g). \tag{1.17}$$
for some continuous functions
$$h_\alpha : U_\alpha \longrightarrow G.$$
It is clear that the maps defined by (1.17) are equivariant since
$$[\varphi_\alpha^{-1}\psi\varphi'_\alpha(x, g)]g_1 = (x, h_\alpha(x)g)g_1 =$$
$$= (x, h_\alpha(x)gg_1) = \varphi_\alpha^{-1}\psi\varphi'_\alpha(x, gg_1).$$

Hence the map ψ is equivariant with respect to the right actions of the group G on the total spaces E and E'. Conversely, let the map ψ be equivariant with respect to the right actions of the group G on the total spaces E and E'. By Theorem 2, the map $\varphi_\alpha^{-1}\psi\varphi_\alpha'$ is equivariant with respect to right translations of the second coordinate of the space $U_\alpha \times G$. Since the map $\varphi_\alpha^{-1}\psi\varphi_\alpha'$ is fiberwise, it has the following form

$$\varphi_\alpha^{-1}\psi\varphi_\alpha'(x,g) = (x, A_\alpha(x,g)). \tag{1.18}$$

The equivariance of the map (1.18) implies that

$$A_\alpha(x, gg_1) = A_\alpha(x, g)g_1$$

for any $x \in U_\alpha$, $g, g_1 \in G$. In particular, putting $g = e$ that

$$A_\alpha(x, g_1) = A_\alpha(x, e)g_1$$

So putting

$$h_\alpha(x) = A_\alpha(x, e),$$

it follows that

$$h_\alpha(x)g = A_\alpha(x, g).$$

and

$$\varphi_\alpha^{-1}\psi\varphi_\alpha'(x,g) = (x, h_\alpha(x)g).$$

The last identity means that ψ is compatible with the structure group G. ∎

Thus by Theorem 3, to show that two locally trivial bundles with the structure group G (and the same base B) are isomorphic it necessary and sufficient to show that there exists an equivariant map of corresponding principal G-bundles (inducing the identity map on the base B). In particular, if one of the bundles is trivial, for instance, $E' = B \times G$, then to construct an equivariant map $\psi : E' \longrightarrow E$ it is sufficient to define a continuous map ψ on the subspace $\{(x, e) : x \in B,\} \subset E' = B \times G$ into E. Then using equivariance, the map ψ is extended by formula

$$\psi(x, g) = \psi(x, e)g.$$

The map $\{(x, e) : x \in B,\} \xrightarrow{\psi} E'$ can be considered as a map

$$s : B \longrightarrow E \tag{1.19}$$

satisfying the property

$$ps(x) = x, \ x \in B. \tag{1.20}$$

The map (1.19) with the property (1.20) is called a *cross-section* of the bundle. So each trivial principal bundle has cross-sections. For instance, the map $B \longrightarrow B \times G$ defined by $x \longrightarrow (x, e)$ is a cross–section. Conversely, if a principal bundle has a cross–section s then this bundle is isomorphic to the trivial principal bundle. The corresponding isomorphism $\psi : B \times G \longrightarrow E$ is defined by formula

$$\psi(x, g) = s(x)g, \quad x \in B, \ g \in G.$$

Let us relax our restrictions on equivariant mappings of principal bundles with the structure group G. Consider arbitrary equivariant mappings of total spaces of principal G–bundles with arbitrary bases. Each fiber of a principal G–bundle is an orbit of the right action of the group G on the total space and hence for each equivariant mapping

$$\psi : E' \longrightarrow E$$

of total spaces, each fiber of the bundle

$$p' : E' \longrightarrow B'$$

maps to a fiber of the bundle

$$p : E \longrightarrow B. \tag{1.21}$$

In other words, the mapping ψ induces a mapping of bases

$$\chi : B' \longrightarrow B \tag{1.22}$$

and the following diagram is commutative

$$
\begin{array}{ccc}
E' & \xrightarrow{\psi} & E \\
\downarrow{\scriptstyle p'} & & \downarrow{\scriptstyle p} \\
B' & \xrightarrow{\chi} & B
\end{array}
\tag{1.23}
$$

Let $U_\alpha \subset B$ be a chart in the base B and let U'_β be a chart such that

$$\chi(U'_\beta) \subset U_\alpha.$$

The mapping $\varphi_\alpha^{-1}\psi\varphi'_\beta$ makes the following diagram commutative

$$
\begin{array}{ccc}
U'_\beta \times G & \xrightarrow{\varphi_\alpha^{-1}\psi\varphi'_\beta} & U_\alpha \times G \\
\downarrow{\scriptstyle p'\varphi'_\beta} & & \downarrow{\scriptstyle p\varphi_\alpha} \\
U'_\beta & \xrightarrow{\chi} & U_\alpha
\end{array}
\tag{1.24}
$$

In diagram (1.24), the mappings $p'\varphi'_\beta$ and $p\varphi_\alpha$ are projections onto the first factors. So one has

$$\varphi_\alpha^{-1}\psi\varphi'_\beta(x', g) = (\chi(x'), h_\beta(x')g).$$

Hence the mapping (1.22) is continuous. Compare the transition functions of these two bundles. First we have

$$(x', \bar{\varphi}'_{\beta_1\beta_2}(x')g) = \varphi'_{\beta_1\beta_2}(x', g) = \varphi'^{-1}_{\beta_1}\varphi'_{\beta_2}(x', g).$$

Then

$$(\chi(x'), h_{\beta_1}(x')\bar{\varphi}'_{\beta_1\beta_2}(x')g) =$$
$$\varphi^{-1}_{\alpha_1}\psi\varphi'_{\beta_1}\varphi'^{-1}_{\beta_1}\varphi'_{\beta_2}(x', g) = \varphi^{-1}_{\alpha_1}\psi\varphi'_{\beta_2}(x', g) =$$
$$= \varphi^{-1}_{\alpha_1}\varphi_{\alpha_1}\varphi^{-1}_{\alpha_2}\psi\varphi'_{\beta_2}(x', g) = (\chi(x'), \bar{\varphi}_{\alpha_1\alpha_2}(\chi(x'))h_{\beta_2}(x')g),$$

that is,

$$h_{\beta_1}(x')\bar{\varphi}_{\beta_1\beta_2}(x') = \bar{\varphi}_{\alpha_1\alpha_2}(\chi(x'))h_{\beta_2}(x'),$$

or

$$h_{\beta_1}(x')\bar{\varphi}'_{\beta_1\beta_2}(x')h^{-1}_{\beta_2}(x') = \bar{\varphi}_{\alpha_1\alpha_2}(\chi(x')). \tag{1.25}$$

By Theorem 1 the left part of (1.25) are the transition functions of a bundle isomorphic to the bundle

$$p' : E' \longrightarrow B'. \tag{1.26}$$

Thus any equivariant mapping of total spaces induces a mapping of bases

$$\chi : B' \longrightarrow B.$$

Moreover, under a proper choice of the coordinate homeomorphisms the transition functions of the bundle (1.26) are inverse images of the transition functions of the bundle (1.21). The inverse is true as well: if

$$\chi : B' \longrightarrow B$$

is a continuous mapping and

$$p : E \longrightarrow B$$

is a principal G–bundle then one can put

$$U'_\alpha = \chi^{-1}(U_\alpha), \quad \bar{\varphi}'_{\alpha\beta}(x') = \bar{\varphi}_{\alpha\beta}(\chi(x)). \tag{1.27}$$

Then the transition functions (1.27) define a principal G–bundle

$$p' : \longrightarrow B,$$

for which there exists an equivariant mapping

$$\psi : E' \longrightarrow E$$

with commutative diagram (1.23). The bundle defined by the transition functions (1.27) is called the *inverse image of the bundle* $p : E \longrightarrow B$ *with respect to the mapping* χ . In the special case when the mapping $\chi : B' \longrightarrow B$ is an inclusion then we say that the inverse image of the bundle with respect to the mapping χ is the *restriction of the bundle to the subspace* $B'' = \chi(B')$. In this case the total space of the restriction of the bundle to the subspace B'' coincides with the inverse image

$$E'' = p^{-1}(B'') \subset E.$$

Thus if

$$E' \xrightarrow{\psi} E$$

is an equivariant mapping of total spaces of principal G–bundles then the bundle $p' : E' \longrightarrow B'$ is an inverse image of the bundle $p : \longrightarrow B$ with respect to the mapping $\chi : B' \longrightarrow B$. Constructing of the inverse image is an important way of construction new locally trivial bundles. The following theorem shows that inverse images with respect to homotopic mappings are isomorphic bundles.

Theorem 4 *Let*

$$p : E \longrightarrow B \times I$$

be a principal G–bundle, where the base is a Cartesian product of the compact space B and the unit interval $I = [0,1]$, and let G be a Lie group. Then restrictions of the bundle p to the subspaces $B \times \{0\}$ and $B \times \{1\}$ are isomorphic.

Proof.

Without of the loss of generality we can assume that the atlas $\{U_\alpha\}$ consists of an atlas $\{V_\beta\}$ on the space B and a finite system of intervals $[a_k, a_{k+1}]$ which cover the segment I, that is,

$$U_\alpha = V_\beta \times [a_k, a_{k+1}].$$

Further, it suffices to assume that there is only one interval $[a_k, a_{k+1}]$ which equals I, so

$$U_\alpha = V_\alpha \times I.$$

Then the transition functions $\varphi_{\alpha\beta}$ depend on the two arguments $x \in V_\alpha \cap V_\beta$ and $t \in I$. Further, we can assume that the transition functions $\varphi_{\alpha\beta}$ are defined and continuous on the closures $\bar{V}_\alpha \cap \bar{V}_\beta$ and thus are uniformly continuous. Hence, we can assume that there exists an open neighborhood O of the neutral element of the group G homeomorphic to a disc and such that

$$\varphi_{\alpha\beta}(x, t_1)\varphi_{\alpha\beta}^{-1}(x, t_2) \in O, \ x \in V_\alpha \cap V_\beta, \ t_1, t_2 \in I,$$

and the power O^N lies in a disk, where N is the number of charts. We construct the functions

$$h_\alpha : V_\alpha \longrightarrow G, \ 1 \leq \alpha \leq N,$$

by induction. Let

$$h_1(x) \equiv 1.$$

Then the function $h_2(x)$ is defined on the set $\bar{V}_1 \cap \bar{V}_2$ by the formula

$$h_2(x) = \varphi_{12}^{-1}(x, 1)\varphi_{12}(x, 0) \in O. \tag{1.28}$$

Then the function (1.28) can be extended on the whole chart V_2, $h_2 : V_2 \longrightarrow O$. Further, notice that when $\alpha < \beta$ the function $h_\beta(x)$ should satisfy the following condition

$$h_\beta(x) = \varphi_{\alpha\beta}^{-1}(x, 1)h_\alpha(x)\varphi_{\alpha\beta}(x, 0). \tag{1.29}$$

on the set $\bar{V}_\alpha \cap \bar{V}_\beta$. Assume that the functions $h_\alpha(x) \in O^\alpha$ are defined for all $\alpha < \beta$ and satisfy the condition (1.29). Then the function $h_\beta(x)$ is well defined on the set

$$\bigcup_{\alpha < \beta} (\bar{V}_\alpha \cap \bar{V}_\beta)$$

and takes values in the set O^β. Hence, the function h_β can be extended on the chart $\bar{V}_\beta$ taking values in O^β. ∎

Corollary 1 *If the transition functions $\varphi_{\alpha\beta}(x)$ and $\psi_{\alpha\beta}(x)$ are homotopic within the class of the transition functions then corresponding bundles are isomorphic.*

Examples

1. In section 1 we considered the Möbius band. The transition functions $\varphi_{\alpha\beta}$ take two values in the homeomorphism group of the fiber: the identity homeomorphism $e(y) \equiv y$, $y \in I$ and homeomorphism $j(y) \equiv 1 - y$, $y \in I$. The group generated by the two elements e and j has the order two since $j^2 = e$. So instead of the Möbius band we can consider corresponding principal bundle with the structure group $G = Z_2$. As a topological space the group G consists of two isolated points. So the fiber of the principal bundle is the discrete two-point space. This fiber space can be thought of as two segments with ends which are identified crosswise. Hence the total space is also a circle and the projection $p : S^1 \longrightarrow S^1$ is a two-sheeted covering. This bundle is nontrivial since the total space of a trivial bundle would have two connected components.

2. In example 3 of section 1 the tangent bundle of two dimensional sphere was considered. The coordinate homeomorphisms

$$\varphi : U \times \mathbf{R}^2 \longrightarrow \mathbf{R}^3 \times \mathbf{R}^3$$

were defined by formulas that were linear with respect to the second argument. Hence the transition functions also have values in the group of linear transformations of the fiber $F = \mathbf{R}^2$, that is, $G = \mathbf{GL}(2, \mathbf{R})$. It can be shown that the structure group G can be reduced to the subgroup $\mathbf{O}(n)$ of orthonormal transformations, induced by rotations and reflections of the plane. Let us explain these statements about the example of the tangent bundle of the two dimensional sphere $\mathbf{S}^2$. To define a coordinate homeomorphism means to define a basis of tangent vectors $e_1(x)$, $e_2(x)$ at each point $x \in U_\alpha$ such that functions $e_1(x)$ and $e_2(x)$ are continuous.

Let us choose two charts $U_\alpha = \{(x, y, z) : z \neq 1\}$, $U_\beta = \{(x, y, z) : z \neq -1\}$. The south pole $P_0 = (0, 0, -1)$ belongs to the chart U_α. The north pole $P_1 = (0, 0, +1)$ belongs to the chart U_β. Consider the meridians. Choose an orthonormal basis for the tangent space of the point P_0 and continue it along the meridians by parallel transfer with respect to the Riemannian metric of the sphere $\mathbf{S}^2$ to all of the chart U_α. Thus we obtain a continuous family of orthonormal bases $e_1(x)$, $e_2(x)$ defined at each point of U_α. In a similar way we construct a continuous family of orthonormal bases $e_1'(x)$, $e_2'(x)$ defined over U_β. Then the coordinate homeomorphisms are defined by the following formulas

$$\begin{aligned}
\varphi_\alpha(x, \xi, \eta) &= \xi e_1(x) + \eta e_2(x) \\
\varphi_\beta(x, \xi', \eta') &= \xi' e_1'(x) + \eta' e_2'(x).
\end{aligned}$$

The transition function $f_{\beta\alpha} = \varphi_\beta^{-1} \varphi_\alpha$ expresses the coordinates of a tangent vector at a point $x \in U_\alpha \cap U_\beta$ in terms of the basis $e_1'(x), e_2'(x)$ by the coordinates of the same vector with respect to the basis $e_1(x), e_2(x)$. As both bases are orthonormal, the change of coordinates (ξ', η') into coordinates (ξ, η) is realized by multiplication by an orthogonal matrix. Thus the structure group $\mathbf{GL}(2, \mathbf{R})$ of the tangent bundle of the sphere $\S^2$ is reducing to the subgroup $\mathbf{O}(2) \subset \mathbf{GL}(2, \mathbf{R})$.

3. Any trivial bundle with the base B can be constructed as the inverse image of the mapping of the base B into a one-point space $\{\mathrm{pt}\}$ which is the base of a trivial bundle.

1.3   VECTOR BUNDLES

The most important special class of locally trivial bundles with given structure group is the class of bundles where the fiber is a vector space and the structure group is a group of linear automorphisms of the vector space. Such bundles are called *vector bundles* . So, for example, the tangent bundle of two-dimensional sphere $\mathbf{S}^2$ is a vector bundle. One can also consider locally trivial bundles where fiber is a infinite dimensional Banach space and the structure group is the group of invertible bounded operators of the Banach space. In the case when the fiber is $\mathbf{R}^n$, the vector bundle ξ is said to be *finite dimensional* and the dimension of the vector bundle is equal to n ($\dim \xi = n$). When the fiber is an infinite dimensional Banach space, the bundle is said to be *infinite dimensional* . Vector bundles possess some special features. First of all notice that each fiber $p^{-1}(x)$, $x \in B$ has the structure of vector space which does not depend on the choice of coordinate homeomorphism. In other words, the operations of addition and multiplication by scalars is independent of the choice of coordinate homeomorphism. Indeed, since the structure group G is $\mathbf{GL}\,(n, \mathbf{R})$ the transition functions

$$\varphi_{\alpha\beta} : (U_\alpha \cap U_\beta) \times \mathbf{R}^n \longrightarrow (U_\alpha \cap U_\beta) \times \mathbf{R}^n$$

are linear mappings with respect to the second factor. Hence a linear combination of vectors goes to the linear combination of images with the same coefficients.

Denote by $\Gamma(\xi)$ the set of all sections of the vector bundle ξ. Then the set $\Gamma(\xi)$ becomes an (infinite dimensional) vector space. To define the structure of vector space on the $\Gamma(\xi)$ consider two sections s_1, s_2:

$$s_1, s_2 : B \longrightarrow E.$$

Put

$$(s_1 + s_2)(x) = s_1(x) + s_2(x), \quad x \in B, \tag{1.30}$$

$$(\lambda s_1)(x) = \lambda(s_1(x)), \quad \lambda \in R, \ x \in B. \tag{1.31}$$

The formulas (1.30) and (1.31) define on the set $\Gamma(\xi)$ the structure of vector space. Notice that an arbitrary section $s : B \longrightarrow E$ can be described in local terms. Let $\{U_\alpha\}$ be an atlas, $\varphi_\alpha : U_\alpha \times \mathbf{R}^n \longrightarrow p^{-1}(U_\alpha)$ be coordinate homeomorphisms, $\varphi_{\alpha\beta} = \varphi_\beta^{-1} \varphi_\alpha$. Then the compositions

$$\varphi_\alpha^{-1} s : U_\alpha \longrightarrow U_\alpha \times \mathbf{R}^n$$

are sections of trivial bundles over U_α and determine vector valued functions $s_\alpha : U_\alpha \longrightarrow \mathbf{R}^n$ by the formula

$$(\varphi_\alpha^{-1})(x) = (x, s_\alpha), \ x \in U_\alpha.$$

On the intersection of two charts $U_\alpha \cap U_\beta$ the functions $s_\alpha(x)$ satisfy the following compatibility condition

$$s_\beta(x) = \varphi_{\beta\alpha}(x)(s_\alpha(x)). \tag{1.32}$$

Conversely, if one has a family of vector valued functions $s_\alpha : U_\alpha \longrightarrow \mathbf{R}^n$ which satisfy the compatibility condition (1.32) then the formula

$$s(x) = \varphi_\alpha(x, s_\alpha(x))$$

determines the mapping $s : B \longrightarrow E$ uniquely (that is, independent of the choice of chart U_α).

The map s is a section of the bundle ξ.

1.3.1 Operations of direct sum and tensor product

There are natural operations induced by the direct sum and tensor product of vector spaces on the family of vector bundles over a common base B. Firstly, consider the operation of direct sum of vector bundles. Let ξ_1 and ξ_2 be two vector bundles with fibers V_1 and V_2, respectively. Denote the transition functions of these bundles in a common atlas of charts by $\varphi_{\alpha\beta}^1(x)$ and $\varphi_{\alpha\beta}^2(x)$. Notice that values of the transition function $\varphi_{\alpha\beta}^1(x)$ lie in the group $\mathbf{GL}\,(V_1)$ whereas the values of the transition function $\varphi_{\alpha\beta}^2(x)$ lie in the group $\mathbf{GL}\,(V_2)$. Hence the transition functions $\varphi_{\alpha\beta}^1(x)$ and $\varphi_{\alpha\beta}^2(x)$ can be considered as matrix–values functions of orders $n_1 = \dim V_1$ and $n_2 = \dim V_2$, respectively. Both of them should satisfy the conditions (1.9) from the section 2.

We form a new space $V = V_1 \oplus V_2$. The linear transformation group $\mathbf{GL}\,(V)$ is the group of matrices of order $n = n_1 + n_2$ which can be decomposed into blocks with respect to decomposition of the space V into the direct sum $V_1 \oplus V_2$. Then the group $\mathbf{GL}\,(V)$ has the subgroup $\mathbf{GL}\,(V_1) \oplus \mathbf{GL}\,(V_2)$ of matrices which have the following form:

$$A = \left\| \begin{array}{cc} A_1 & 0 \\ 0 & A_2 \end{array} \right\| = A_1 \oplus A_2, \ A_1 = \mathbf{GL}\,(V_1), \ A_2 = \mathbf{GL}\,(V_2).$$

Then we can construct new the transition functions

$$\varphi_{\alpha\beta}(x) = \varphi_{\alpha\beta}^1(x) \oplus \varphi_{\alpha\beta}^2(x) = \left\| \begin{matrix} \varphi_{\alpha\beta}^1(x) & 0 \\ 0 & \varphi_{\alpha\beta}^2(x) \end{matrix} \right\|. \qquad (1.33)$$

The transition functions (1.33) satisfy the conditions (1.9) from section 2, that is, they define a vector bundle with fiber $V = V_1 \oplus V_2$. The bundle constructed above is called the *direct sum of vector bundles* ξ_1 and ξ_2 and is denoted by $\xi = \xi_1 \oplus \xi_2$. The direct sum operation can be constructed in a geometric way. Namely, let $p_1 : E_1 \longrightarrow B$ be a vector bundle ξ_1 and let $p_2 : E_2 \longrightarrow B$ be a vector bundle ξ_2. Consider the Cartesian product of total spaces $E_1 \times E_2$ and the projection

$$p_3 = p_1 \times p_2 : E_1 \times E_2 \longrightarrow B \times B.$$

It is clear that p is vector bundle with the fiber $V = V_1 \oplus V_2$.

Consider the diagonal $\Delta \subset B \times B$, that is, the subset $\Delta = \{(x, x) : x \in B\}$. The diagonal Δ is canonically homeomorphic to the space B. The restriction of the bundle p_3 to $\Delta \approx B$ is a vector bundle over B. The total space E of this bundle is the subspace $E \subset E_1 \times E_2$ that consists of the vectors (y_1, y_2) such that

$$p_1(y_1) = p_2(y_2).$$

It is easy to check that $\{U_{\alpha_1} \times U_{\alpha_2}\}$ gives an atlas of charts for the bundle p_3.

The transition functions $\varphi_{(\beta_1\beta_2)(\alpha_1\alpha_2)}(x, y)$ on the intersection of two charts $(U_{\alpha_1} \times U_{\alpha_2}) \cap (U_{\beta_1} \times U_{\beta_2})$ have the following form:

$$\varphi_{(\beta_1\beta_2)(\alpha_1\alpha_2)}(x, y) = \left\| \begin{matrix} \varphi_{\beta_1\alpha_1}^1(x) & 0 \\ 0 & \varphi_{\beta_2\alpha_2}^2(y) \end{matrix} \right\|.$$

Hence on the diagonal $\Delta \approx B$ the atlas consists of sets $U_\alpha \approx \Delta \cap (U_\alpha \times U_\alpha)$.

Then the transition functions for the restriction of the bundle p_3 on the diagonal have the following form:

$$\varphi_{(\beta\beta)(\alpha\alpha)}(x, x) = \left\| \begin{matrix} \varphi_{\beta\alpha}^1(x) & 0 \\ 0 & \varphi_{\beta\alpha}^2(x) \end{matrix} \right\|. \qquad (1.34)$$

So the transition functions (1.34) coincide with the transition functions defined for the direct sum of the bundles ξ_1 and ξ_2. Now let us proceed to the definition of tensor product of vector bundles. As before, let ξ_1 and ξ_2 be two vector bundles with fibers V_1 and V_2 and let $\varphi_{\alpha\beta}^1(x)$ and $\varphi_{\alpha\beta}^2(x)$ be the transition functions of the vector bundles ξ_1 and ξ_2,

$$\varphi_{\alpha\beta}^1(x) \in \mathbf{GL}\,(V_1),\ \varphi_{\alpha\beta}^2(x) \in \mathbf{GL}\,(V_2),\ x \in V_\alpha \cap V_\beta.$$

Let $V = V_1 \otimes V_2$. Then form the tensor product $A_1 \otimes A_2 \in \mathbf{GL}\,(V_1 \otimes V_2)$ of the two matrices $A_1 \in \mathbf{GL}\,(V_1)$, $A_2 \in \mathbf{GL}\,(V_2)$. Put

$$\varphi_{\alpha\beta}(x) = \varphi_{\alpha\beta}^1(x) \otimes \varphi_{\alpha\beta}^2(x).$$

Now we have obtained a family of the matrix value functions $\varphi_{\alpha\beta}(x)$ which satisfy the conditions (1.9) from the section 2. The corresponding vector bundle ξ with fiber $V = V_1 \otimes V_2$ and transition functions $\varphi_{\alpha\beta}(x)$ will be called *the tensor product* of bundles ξ_1 and ξ_2 and denoted by

$$\xi = \xi_1 \otimes \xi_2.$$

What is common in the construction of the operations of direct sum and operation of tensor product? Both operations can be described as the result of applying the following sequence of operations to the pair of vector bundles ξ_1 and ξ_2:

1. Pass to the principal $\mathbf{GL}\,(V_1)$– and $\mathbf{GL}\,(V_2)$– bundles;

2. Construct the principal ($\mathbf{GL}\,(V_1) \times \mathbf{GL}\,(V_2)$)– bundle over the Cartesian square $B \times B$;

3. Restrict to the diagonal Δ, homeomorphic to the space B.

4. Finally, form a new principal bundle by means of the relevant representations of the structure group $\mathbf{GL}\,(V_1) \times \mathbf{GL}\,(V_2)$ in the groups $\mathbf{GL}\,(V_1 \oplus V_2)$ and $\mathbf{GL}\,(V_1 \otimes V_2)$, respectively.

The difference between the operations of direct sum and tensor product lies in choice of the representation of the group $\mathbf{GL}\,(V_1) \times \mathbf{GL}\,(V_2)$.

By using different representations of the structure groups, further operations of vector bundles can be constructed, and algebraic relations holding for representations induce corresponding algebraic relations vector bundles.

In particular, for the operations of direct sum and tensor product the following well known relations hold:

1. Associativity of the direct sum

$$(\xi_1 \oplus \xi_2) \oplus \xi_3 = \xi_1 \oplus (\xi_2 \oplus \xi_3).$$

This relation is a consequence of the commutative diagram

$$\mathbf{GL}\ (V_1 \oplus V_2) \times \mathbf{GL}\ (V_3)$$

$$\nearrow \rho_1 \qquad \searrow \rho_2$$

$$\mathbf{GL}\ (V_1) \times \mathbf{GL}\ (V_2) \times \mathbf{GL}\ (V_3) \qquad \mathbf{GL}\ (V_1 \oplus V_2 \oplus V_3)$$

$$\searrow \rho_3 \qquad \nearrow \rho_4$$

$$\mathbf{GL}\ (V_1) \times \mathbf{GL}\ (V_2 \oplus V_3)$$

where

$$\begin{aligned}
\rho_1(A_1, A_2, A_3) &= (A_1 \oplus A_2, A_3), \\
\rho_2(B, A_3) &= B \oplus A_3, \\
\rho_3(A_1, A_2, A_3) &= (A_1, A_2 \oplus A_3), \\
\rho_4(A_1, C) &= A_1 \oplus C.
\end{aligned}$$

Then

$$\begin{aligned}
\rho_2\rho_1(A_1, A_2, A_3) &= (A_1 \oplus A_2) \oplus A_3, \\
\rho_4\rho_3(A_1, A_2, A_3) &= A_1 \oplus (A_2 \oplus A_3).
\end{aligned}$$

It is clear that

$$\rho_2\rho_1 = \rho_4\rho_3$$

since the relation

$$(A_1 \oplus A_2) \oplus A_3 = A_1 \oplus (A_2 \oplus A_3)$$

is true for matrices.

2. Associativity for tensor products

$$(\xi_1 \otimes \xi_2) \otimes \xi_3 = \xi_1 \otimes (\xi_2 \otimes \xi_3).$$

This relation is a consequence of the commutative diagram

$$\mathbf{GL}\ (V_1 \otimes V_2) \times \mathbf{GL}\ (V_3)$$

$$\nearrow \rho_1 \qquad \searrow \rho_2$$

$$\mathbf{GL}\ (V_1) \times \mathbf{GL}\ (V_2) \times \mathbf{GL}\ (V_3) \qquad \mathbf{GL}\ (V_1 \otimes V_2 \otimes V_3)$$

$$\searrow \rho_3 \qquad \nearrow \rho_4$$

$$\mathbf{GL}\ (V_1) \times \mathbf{GL}\ (V_2 \otimes V_3)$$

The commutativity is implied from the relation

$$(A_1 \otimes A_2) \otimes A_3 = A_1 \otimes (A_2 \otimes A_3)$$

for matrices.

3. Distributivity:

$$(\xi_1 \oplus \xi_2) \otimes \xi_3 = (\xi_1 \otimes \xi_3) \oplus (\xi_2 \otimes \xi_3).$$

This property is implied by the corresponding relation

$$(A_1 \oplus A_2) \otimes A_3 = (A_1 \otimes A_3) \oplus (A_2 \otimes A_3).$$

for matrices.

4. Denote the trivial vector bundle with the fiber $\mathbf{R}^n$ by $\bar{n}$. The total space of trivial bundle is homeomorphic to the Cartesian product $B \times R_n$ and it follows that

$$\bar{n} = \bar{1} \oplus \bar{1} \oplus \ldots \oplus \bar{1}(n \text{ times}).$$

and

$$\xi \otimes \bar{1} = \xi,$$

$$\xi \otimes \bar{n} = \xi \oplus \xi \oplus \ldots \oplus \xi(n \text{ times}).$$

1.3.2   Other operations with vector bundles

Let $V = \mathbf{Hom}\,(V_1, V_2)$ be the vector space of all linear mappings from the space V_1 to the space V_2. For infinite dimensional Banach spaces we will assume that all linear mappings considered are bounded. Then there is a natural representation of the group $\mathbf{GL}\,(V_1) \times \mathbf{GL}\,(V_2)$ into the group $\mathbf{GL}\,(V)$ which to any pair $A_1 \in \mathbf{GL}\,(V_1)$, $A_2 \in \mathbf{GL}\,(V_2)$ associates the mapping

$$\rho(A_1, A_2) : \mathbf{Hom}\,(V_1, V_2) \longrightarrow \mathbf{Hom}\,(V_1, V_2)$$

by the formula

$$\rho(A_1, A_2)(f) = A_2 \circ f \circ A_1^{-1}. \tag{1.35}$$

Then following the general method of constructing operations for vector bundles one obtains for each pair of vector bundles ξ_1 and ξ_2 with fibers V_1 and V_2 and transition functions $\varphi_{\alpha\beta}^1(x)$ and $\varphi_{\alpha\beta}^2(x)$ a new vector bundle with fiber V and transition functions

$$\varphi_{\alpha\beta}(x) = \rho(\varphi_{\alpha\beta}^1(x), \varphi_{\alpha\beta}^2(x)) \tag{1.36}$$

This bundle is denoted by $\mathbf{HOM}\,(\xi_1, \xi_2)$.

When $V_2 = \mathbf{R}^1$, the space $\mathbf{Hom}\,(V_1, R^1)$ is denoted by V_1^*. Correspondingly, when $\xi_2 = \bar{1}$ the bundle $\mathbf{HOM}\,(\xi, \bar{1})$ will be denoted by ξ^* and called the *dual* *bundle* . It is easy to check that the bundle ξ^* can be constructed from ξ by means of the representation of the group $\mathbf{GL}\,(V)$ to itself by the formula

$$A \longrightarrow (A^t)^{-1},\ A \in \mathbf{GL}\,(V).$$

There is a bilinear mapping

$$V \times V^* \overset{\beta}{\longrightarrow} \mathbf{R}^1,$$

which to each pair (x, h) associates the value $h(x)$.

Consider the representation of the group $\mathbf{GL}\,(V)$ on the space $V \times V^*$ defined by matrix

$$A \longrightarrow \left\| \begin{array}{cc} A & 0 \\ 0 & \rho(A, 1) \end{array} \right\|$$

(see (1.35)). Then the structure group $\mathbf{GL}\,(V \times V^*)$ of the bundle $\xi \oplus \xi^*$ is reduced to the subgroup $\mathbf{GL}\,(V)$. The action of the group $\mathbf{GL}\,(V)$ on the $V \times V^*$ has the property that the mapping β is equivariant with respect to trivial action of the group $\mathbf{GL}\,(V)$ on $\mathbf{R}^1$. This fact means that the value of the form h on the vector x does not depend on the choice of the coordinate system in the space V. Hence there exists a continuous mapping

$$\bar{\beta} : \xi \oplus \beta^* \longrightarrow \bar{1},$$

which coincides with β on each fiber.

Let $\Lambda_k(V)$ be the k-th exterior power of the vector space V. Then to each transformation $A : V \longrightarrow V$ is associated the corresponding exterior power of the transformation

$$\Lambda_k(A) : \Lambda_k(V) \longrightarrow \Lambda_k(V),$$

that is, the representation

$$\Lambda_k : \mathbf{GL}\,(V) \longrightarrow \mathbf{GL}\,(\Lambda_k(V)).$$

The corresponding operation for vector bundles will be called *the operation of the k-th exterior power* and the result denoted by $\Lambda_k(\xi)$. Similar to vector spaces, for vector bundles one has

$$\Lambda_1(\xi) = \xi,$$

$$\Lambda_k(\xi) = 0 \text{ for } k > \dim \xi,$$

$$\Lambda_k(\xi_1 \oplus \xi_2) = \oplus_{\alpha=0}^{k} \Lambda_\alpha(\xi_1) \otimes \Lambda_{k-\alpha}(\xi_2), \tag{1.37}$$

where by definition
$$\Lambda_0(\xi) = \bar{1}.$$
It is convenient to write the relation (1.37) using the partition function. Let us introduce the polynomial
$$\Lambda_t(\xi) = \Lambda_0(\xi) + \Lambda_1(\xi)t + \Lambda_2(\xi)t^2 + \ldots + \Lambda_n(\xi)t^n.$$
Then
$$\Lambda_t(\xi_1 \oplus \xi_2) = \Lambda_t(\xi_1) \otimes \Lambda_t(\xi_2). \qquad (1.38)$$
and the formula (1.38) should be interpreted as follows: the degrees of the formal variable are added and the coefficients are vector bundles formed using the operations of tensor product and direct sum.

1.3.3 Mappings of vector bundles

Consider two vector bundles ξ_1 and ξ_2 where
$$\xi_i = \{p_i : E_i \longrightarrow B, \; V_i \text{ is fiber}\}.$$
Consider a fiberwise continuous mapping
$$f : E_1 \longrightarrow E_2.$$
The map f will be called a *linear map of vector bundles* or *homomorphism of bundles* if f is linear on each fiber. The family of all such linear mappings will be denoted by **Hom** (ξ_1, ξ_2). Then the following relation holds:
$$\textbf{Hom } (\xi_1, \xi_2) = \Gamma(\textbf{HOM } (\xi_1, \xi_2)). \qquad (1.39)$$

By intuition, the relation (1.39) is evident since elements from both the left-hand and right-hand sides are families of linear transformations from the fiber V_1 to the fiber V_2, parametrized by points of the base B.

To prove the relation (1.39), let us express elements from both the left-hand and right-hand sides of (1.39) in terms of local coordinates. Consider an atlas $\{U_\alpha\}$ and coordinate homeomorphisms φ^1_α, φ^2_α for bundles ξ_1, ξ_2. By means of the mapping $f : E_1 \longrightarrow E_2$ we construct a family of mappings:
$$(\varphi^2_\alpha)^{-1} f \varphi^1_\alpha : U_\alpha \times V_1 \longrightarrow U_\alpha \times V_2,$$
defined by the formula:
$$[(\varphi^2_\alpha)^{-1} f \varphi^1_\alpha](x, h) = (x, f_\alpha(x)h),$$

for the continuous family of linear mappings

$$f_\alpha(x) : V_1 \longrightarrow V_2.$$

On the intersection of two charts $U_\alpha \cap U_\beta$ two functions $f_\alpha(x)$ and $f_\beta(x)$ satisfy the following condition

$$\varphi_{\beta\alpha}^2(x) f_\alpha(x) = f_\beta(x) \varphi_{\beta\alpha}^1(x),$$

or

$$f_\beta(x) = \varphi_{\beta\alpha}^2(x) f_\alpha(x) \varphi_{\alpha\beta}^1(x).$$

Taking into account the relations (1.35), (1.36) we have

$$f_\beta(x) = \varphi_{\beta\alpha}(x)(f_\alpha(x)). \tag{1.40}$$

In other words, the family of functions

$$f_\alpha(x) \in V = \mathbf{Hom}\ (V_1, V_2),\ x \in U_\alpha$$

satisfies the condition (1.40), that is, determines a section of the bundle $\mathbf{HOM}\ (\xi_1, \xi_2)$. Conversely, given a section of the bundle $\mathbf{HOM}\ (\xi_1, \xi_2)$, that is, a family of functions $f_\alpha(x)$ satisfying condition (1.40) defines a linear mapping from the bundle ξ_1 to the bundle ξ_2. In particular, if

$$\xi_1 = \bar{1},\ V_1 = \mathbf{R}^1$$

then

$$\mathbf{Hom}\ (V_1, V_2) = V_2.$$

Hence

$$\mathbf{HOM}\ (\bar{1}, \xi_2) = \xi_2.$$

Hence

$$\Gamma(\xi_2) = \mathbf{Hom}\ (\bar{1}, \xi_2),$$

that is, the space of all sections of vector bundle ξ_2 is identified with the space of all linear mappings from the one dimensional trivial bundle $\bar{1}$ to the bundle ξ_2.

The second example of mappings of vector bundles gives an analogue of bilinear form on vector bundle. A bilinear form is a mapping

$$V \times V \longrightarrow \mathbf{R}^1,$$

which is linear with respect to each argument. Consider a continuous family of bilinear forms parametrized by points of base. This gives us a definition of bilinear form on vector bundle, namely, a fiberwise continuous mapping

$$f : \xi \oplus \xi \longrightarrow \bar{1} \tag{1.41}$$

which is bilinear in each fiber and is called a *bilinear form on the bundle* ξ. Just as on a linear space, a bilinear form on a vector bundle (1.41) induces a linear mapping from the vector bundle ξ to its dual bundle ξ^*

$$\bar{f} : \xi \longrightarrow \xi^*,$$

such that f decomposes into the composition

$$\xi \oplus \xi \overset{\bar{f} \oplus \mathbf{Id}}{\longrightarrow} \xi^* \oplus \xi \overset{\beta}{\longrightarrow} \bar{1},$$

where

$$\mathbf{Id} : \xi \longrightarrow \xi$$

is the identity mapping and

$$\xi \oplus \xi \overset{\bar{f} \oplus \mathbf{Id}}{\longrightarrow} \xi^* \oplus \xi$$

is the direct sum of mappings $\bar{f}$ and $\mathbf{Id}$ on each fiber. When the bilinear form f is symmetric, positive and nondegenerate we say that f is a *scalar product on the bundle* ξ.

Theorem 5 *Let ξ be a finite dimensional vector bundle over a compact base space B. Then there exists a scalar product on the bundle ξ, that is, a nondegenerate, positive, symmetric bilinear form on the ξ.*

Proof.

We must construct a fiberwise mapping (1.41) which is bilinear, symmetric, positive, nondegenerate form in each fiber. This means that if $x \in B$, $v_1, v_2 \in p^{-1}(x)$ then the value $f(v_1, v_2)$ can be identified with a real number such that

$$f(v_1, v_2) = f(v_2, v_1)$$

and

$$f(v, v) > 0 \text{ for any } v \in p^{-1}(x), \ v \neq 0.$$

Consider the weaker condition

$$f(v, v) \geq 0.$$

Then we obtain a nonnegative bilinear form on the bundle ξ. If f_1, f_2 are two nonnegative bilinear forms on the bundle ξ then the sum $f_1 + f_2$ and a linear combination $\varphi_1 f_1 + \varphi_2 f_2$ for any two nonnegative continuous functions φ_1 and φ_2 on the base B gives a nonnegative bilinear form as well.

Let $\{U_\alpha\}$ be an atlas for the bundle ξ. The restriction $\xi|_{U_\alpha}$ is a trivial bundle and is therefore isomorphic to a Cartesian product $U_\alpha \times V$ where V is fiber of ξ. Therefore the bundle $\xi|_{U_\alpha}$ has a nondegenerate positive definite bilinear form

$$f_\alpha : \xi|_{U_\alpha} \oplus \xi|_{U_\alpha} \longrightarrow \bar{1}.$$

In particular, if $v \in p^{-1}(x)$, $x \in U_\alpha$ and $v \neq 0$ then

$$f_\alpha(v, v) > 0.$$

Consider a partition of unity $\{g_\alpha\}$ subordinate to the atlas $\{U_\alpha\}$. Then

$$0 \leq g_\alpha(x) \leq 1,$$

$$\sum_\alpha g_\alpha(x) \equiv 1,$$

$$\textbf{Supp } g_\alpha \subset U_\alpha.$$

We extend the form f_α by formula

$$\bar{f}_\alpha(v_1, v_2) = \begin{cases} g_\alpha(x) f_\alpha(v_1, v_2) & v_1, v_2 \in p^{-1}(x) \quad x \in U_\alpha, \\ 0 & v_1, v_2 \in p^{-1}(x) \quad x \notin U_\alpha. \end{cases} \qquad (1.42)$$

It is clear that the form (1.42) defines a continuous nonnegative form on the bundle ξ. Put

$$f(v_1, v_2) = \sum_\alpha f_\alpha(v_1, v_2). \qquad (1.43)$$

The form (1.43) is then positive definite. Actually, let $0 \neq v \in p^{-1}(x)$. Then there is an index α such that

$$g_\alpha(x) > 0.$$

This means that

$$x \in U_\alpha \text{ and } f_\alpha(v, v) > 0.$$

Hence

$$\bar{f}_\alpha(v, v) > 0$$

and

$$f(v, v) > 0.$$

Theorem 6 *For any vector bundle ξ over a compact base space B with $\dim \xi = n$, the structure group $\mathbf{GL}\,(n, \mathbf{R})$ reduces to subgroup $\mathbf{O}(n)$.*

Proof.

Let us give another geometric interpretation of the property that the bundle ξ is locally trivial. Let U_α be a chart and let

$$\varphi_\alpha : U_\alpha \times V \longrightarrow p^{-1}(U_\alpha)$$

be a trivializing coordinate homeomorphism. Then any vector $v \in V$ defines a section of the bundle ξ over the chart U_α

$$\sigma : U_\alpha \longrightarrow p^{-1}(U_\alpha),$$
$$\sigma(x) \;\; = \;\; \varphi_\alpha(x, v) \in p^{-1}(U_\alpha).$$

If $v_1, \ldots, v_n$ is a basis for the space V then corresponding sections

$$\sigma_k^\alpha(x) = \varphi_\alpha(x, v_k)$$

form a system of sections such that for each point $x \in U_\alpha$ the family of vectors $\sigma_1^\alpha(x), \ldots, \sigma_n^\alpha(x) \in p^{-1}(x)$ is a basis in the fiber $p^{-1}(x)$.

Conversely, if the system of sections

$$\sigma_1^\alpha, \ldots, \sigma_n^\alpha : U_\alpha \longrightarrow p^{-1}(U_\alpha)$$

forms basis in each fiber then we can recover a trivializing coordinate homeomorphism

$$\varphi_\alpha\left(x, \sum_i \lambda_i v_i\right) = \sum_i \lambda_i \sigma_i^\alpha(x) \in p^{-1}(U_\alpha).$$

From this point of view, the transition function $\varphi^{\beta\alpha} = \varphi_\beta^{-1}\varphi_\alpha$ has an interpretation as a change of basis matrix from the basis $\{\sigma_1^\alpha(x), \ldots, \sigma_n^\alpha(x)\}$ to $\{\sigma_1^\beta(x), \ldots, \sigma_n^\beta(x)\}$ in the fiber $p^{-1}(x)$, $x \in U_\alpha \cap U_\beta$. Thus Theorem 6 will be proved if we construct in each chart U_α a system of sections $\{\sigma_1^\alpha, \ldots, \sigma_n^\alpha\}$ which form an orthonormal basis in each fiber with respect to a inner product in the bundle ξ. Then the transition matrices from one basis $\{\sigma_1^\alpha(x), \ldots, \sigma_n^\alpha(x)\}$ to another basis $\{\sigma_1^\beta(x), \ldots, \sigma_n^\beta(x)\}$ will be orthonormal, that is, $\varphi_{\beta\alpha}(x) \in \mathbf{O}(n)$. The proof of Theorem 6 will be completed by the following lemma.

Lemma 1 *Let ξ be a vector bundle, f a scalar product in the bundle ξ and $\{U_\alpha\}$ an atlas for the bundle ξ. Then for any chart U_α there is a system of sections $\{\sigma_1^\alpha, \ldots, \sigma_n^\alpha\}$ orthonormal in each fiber $p^{-1}(x)$, $x \in U_\alpha$.*

Proof.

The proof of the lemma simply repeats the Gramm-Schmidt method of construction of orthonormal basis. Let

$$\tau_1, \ldots, \tau_n : U_\alpha \longrightarrow p^{-1}(U_\alpha)$$

be an arbitrary system of sections forming a basis in each fiber $p^{-1}(x)$, $x \in U_\alpha$. Since for any $x \in U_\alpha$,

$$\tau_1(x) \neq 0$$

one has

$$f(\tau_1(x), \tau_1(x)) > 0.$$

Put

$$\tau_1'(x) = \frac{\tau_1(x)}{\sqrt{f(\tau_1(x), \tau_1(x))}}.$$

The new system of sections $\tau_1', \tau_2, \ldots, \tau_n$ forms a basis in each fiber. Put

$$\tau_2''(x) = \tau_2(x) - f(\tau_2(x), \tau_1'(x))\tau_1'(x).$$

The new system of sections $\tau_1', \tau_2'', \tau_3(x), \ldots, \tau_n$ forms a basis in each fiber. The vectors $\tau_1'(x)$ have unit length and are orthogonal to the vectors $\tau_2''(x)$ at each point $x \in U_\alpha$. Put

$$\tau_2'(x) = \frac{\tau_2''(x)}{\sqrt{f(\tau_2''(x), \tau_2''(x))}}.$$

Again, the system of sections $\tau_1', \tau_2', \tau_3(x), \ldots, \tau_n$ forms a basis in each fiber and, moreover, the vectors τ_1', τ_2' are orthonormal.

Then we rebuild the system of sections by induction. Let the sections $\tau_1', \ldots, \tau_k'$, $\tau_{k+1}(x), \ldots, \tau_n$ form a basis in each fiber and suppose that the sections $\tau_1', \ldots, \tau_k'$ be are orthonormal in each fiber. Put

$$\tau_{k+1}''(x) = \tau_{k+1}(x) - \sum_{i=1}^{k} f(\tau_{k+1}(x), \tau_i'(x))\tau_i'(x),$$

$$\tau_{k+1}'(x) = \frac{\tau_{k+1}''(x)}{\sqrt{f(\tau_{k+1}''(x), \tau_{k+1}''(x))}}.$$

It is easy to check that the system $\tau_1', \ldots, \tau_{k+1}', \tau_{k+2}(x), \ldots, \tau_n$ forms a basis in each fiber and the sections $\tau_1', \ldots, \tau_{k+1}'$ are orthonormal. The lemma is proved by induction. Thus the proof ofh Theorem 6 is finished. ∎

Remarks

1. In Lemma 1 we proved a stronger statement: if $\{\tau_1, \ldots, \tau_n\}$ is a system of sections of the bundle ξ in the chart U_α which is a basis in each fiber $p^{-1}(x)$ and if in addition vectors $\{\tau_1, \ldots, \tau_k\}$ are orthonormal then there are sections $\{\tau'_{k+1}, \ldots, \tau'_n\}$ such that the system

$$\{\tau_1, \ldots, \tau_k, \tau'_{k+1}, \ldots, \tau'_n\}$$

is orthonormal in each fiber. In other words, if a system of orthonormal sections can be extended to basis then it can be extended to orthonormal basis.

2. In theorems 5 and 6 the condition of compactness of the base B can be replaced by the condition of paracompactness. In the latter case we should first choose a locally finite atlas of charts.

1.4 LINEAR TRANSFORMATIONS OF BUNDLES

Many properties of linear mappings between vector spaces can be extended to linear mappings or homomorphisms between vector bundles. We shall consider some of them in this section. Fix a vector bundle ξ with the base B equipped with a scalar product. The value of the scalar product on a pair of vectors v_1, $v_2 \in p^{-1}(x)$ will be denoted by $\langle v_1, v_2 \rangle$.

Consider a homomorphism $f : \xi \longrightarrow \xi$ of the vector bundle ξ to itself. The space of all such homomorphisms **Hom** (ξ, ξ) has the natural operations of summation and multiplication by continuous functions. Thus the space **Hom** (ξ, ξ) is a module over the algebra $C(B)$ of continuous functions on the base B. Further, the operation of composition equips the space **Hom** (ξ, ξ) with the structure of algebra. Using a scalar product in the bundle ξ one can introduce a norm in the algebra **Hom** (ξ, ξ) and hence equip it with a structure of a Banach algebra: for each homomorphism $f : \xi \longrightarrow \xi$ put

$$\|f\| = \sup_{v \neq 0} \frac{\|f(v)\|}{\|v\|}, \qquad (1.44)$$

where

$$\|v\| = \sqrt{\langle v, v \rangle}.$$

The space **Hom** (ξ, ξ) is complete with respect to the norm (1.44). Indeed, if a sequence of homomorphisms $f_n : \xi \longrightarrow \xi$ is a Cauchy sequence with respect to this norm, that is,

$$\lim_{n,m \longrightarrow \infty} \|f_n - f_m\| = 0$$

then for any fixed vector $v \in p^{-1}(x)$ the sequence $f_n(v) \in p^{-1}(x)$ is a Cauchy sequence as well since

$$\|f_n(v) - f_m(v)\| leq \|f_n - f_m\| \cdot \|v\|.$$

Hence there exists a limit

$$f(v) = \lim_{n \longrightarrow \infty} f_n(v).$$

The mapping $f : \xi \longrightarrow \xi$ is evidently linear. To show its continuity one should consider the mappings

$$h_{n,\alpha} = \varphi_\alpha^{-1} f_n \varphi_\alpha : U_\alpha \times V \longrightarrow U_\alpha \times V,$$

which are defined by the matrix valued functions on the U_α. The coefficients of these matrices give Cauchy sequences in the uniform norm and therefore the limit values are continuous functions.

This means that f is continuous. The scalar product in the vector bundle ξ defines an adjoint linear mapping f^* by the formula

$$\langle f^*(v_1), v_2 \rangle = \langle v_1, f(v_2) \rangle, \ v_1, v_2 \in p^{-1}(x).$$

The proof of existence and continuity of the homomorphism f^* is left to reader as an exercise.

1.4.1 Complex bundles

The identity homomorphism of the bundle ξ to itself will be denoted by 1. Consider a homomorphism

$$I : \xi \longrightarrow \xi$$

which satisfies the condition

$$I^2 = -1.$$

The restriction

$$I_x = I_{|p^{-1}(x)}$$

is an automorphism of the fiber $p^{-1}(x)$ with the property

$$I_x^2 = -1.$$

Hence the transformation I_x defines a structure of a complex vector space on $p^{-1}(x)$.

In particular, one has

$$\dim V = 2n.$$

Let us show that in this case the structure group $\mathbf{GL}\,(2n, \mathbf{R})$ is reduced to the subgroup of complex transformations $\mathbf{GL}\,(n, \mathbf{C})$.

First of all notice that if the system of vectors $\{v_1, \ldots, v_n\}$ has the property that the system $\{v_1, \ldots, v_n, Iv_1, \ldots, Iv_n\}$ is a real basis in the space V, $\dim V = 2n$, then the system $\{v_1, \ldots, v_n\}$ is a complex basis of V. Fix a point x_0. The space $p^{-1}(x)$ is a complex vector space with respect to the operator I_{x_0}, and hence there is a complex basis $\{v_1, \ldots, v_n\} \subset p^{-1}(x)$.

Let $U_\alpha \ni x_0$ be a chart for the bundle ξ. There are sections $\{\tau_1, \ldots, \tau_n\}$ in the chart U_α such that

$$\tau_k(x_0) = v_k, 1 \le k \le n.$$

Then the system of sections $\{\tau_1, \ldots, \tau_n, I\tau_1, \ldots, I\tau_n\}$ forms a basis in the fiber $p^{-1}(x_0)$ and therefore forms basis in each fiber $p^{-1}(x)$ in sufficiently small neighborhood $U \ni x_0$. Hence the system $\{\tau_1(x), \ldots, \tau_n(x)\}$ forms a complex basis in each fiber $p^{-1}(x)$ in the neighborhood $U \ni x_0$. This means that there is a sufficient fine atlas $\{U_\alpha\}$ and a system of sections $\{\tau_1^\alpha(x), \ldots, \tau_n^\alpha(x)\}$ on each chart U_α giving a complex basis in each fiber $p^{-1}(x)$ in the neighborhood $U_\alpha \ni x_0$. Fix a complex basis $\{e_1, \ldots, e_n\}$ in the complex vector space $\mathbf{C}^n$. Put

$$\varphi_\alpha : U_\alpha \times \mathbf{C}^n \longrightarrow p^{-1}(U_\alpha),$$

$$\varphi_\alpha(x, \sum_{k=1}^{n} z_k e_k) = \sum_{k=1}^{n} u_k \tau_k^\alpha(x) + \sum_{k=1}^{n} v_k I_x(\tau_k^\alpha(x))$$

$$= \sum_{k=1}^{n} z_k \tau_k^\alpha(x),$$

$$z_k = u_k + iv_k, \ 1 \le k \le n.$$

Then the transition functions $\varphi_{\beta\alpha} = \varphi_\beta^{-1}\varphi_\alpha$ are determined by the transition matrices from the complex basis $\{\tau_1^\alpha(x), \ldots, \tau_n^\alpha(x)\}$ to the complex basis $\{\tau_1^\beta(x), \ldots, \tau_n^\beta(x)\}$. These matrices are complex, that is, belong to the group

GL $(n, \mathbf{C})$. A vector bundle with the structure group **GL** $(n, \mathbf{C})$ is called a *complex vector bundle* .

Let ξ be a real vector bundle. In the vector bundle $\xi \oplus \xi$, introduce the structure of a complex vector bundle by means the homomorphism

$$I : \xi \oplus \xi \longrightarrow \xi \oplus \xi$$

given by

$$I(v_1, v_2) = (-v_2, v_1), \ v_1, v_2 \in p^{-1}(x). \tag{1.45}$$

The complex vector bundle defined by (1.45) is called the *complexification of the bundle* ξ and is denoted by $c\xi$. Conversely, forgetting of the structure of complex bundle on the complex vector bundle ξ turns it into a real vector bundle. This operation is called the *realification of a complex vector bundle* ξ and is denoted by $r\xi$. It is clear that

$$rc\xi = \xi \oplus \xi.$$

The operations described above correspond to natural representations of groups:

$$c : \ \mathbf{GL} \ (n, \mathbf{R}) \longrightarrow \mathbf{GL} \ (n, \mathbf{C}),$$

$$r : \ \mathbf{GL} \ (n, \mathbf{C}) \longrightarrow \mathbf{GL} \ (2n, \mathbf{R}).$$

Let us clarify the structure of the bundle $cr\xi$. If ξ is a complex bundle, that is, ξ is a real vector bundle with a homomorphism $I : \xi \longrightarrow \xi$ giving the structure of complex bundle on it. By definition the vector bundle $cr\xi$ is a new real vector bundle $\eta = \xi \oplus \xi$ with the structure of complex vector bundle defined by a homomorphism

$$
\begin{aligned}
I_1 : \xi \oplus \xi \ &\longrightarrow \ \xi \oplus \xi, \\
I_1(v_1, v_2) \ &= \ (-v_2, v_1).
\end{aligned}
\tag{1.46}
$$

The mapping (1.46) defines a new complex structure in the vector bundle η which is, in general, different from the complex structure defined by the mapping I.

Let us split the bundle η in another way:

$$
\begin{aligned}
f : \xi \oplus \xi \ &\longrightarrow \ \xi \oplus \xi, \\
f(v_1, v_2) \ &= \ (I(v_1 + v_2), v_1 - v_2).
\end{aligned}
$$

and define in the inverse image a new homomorphism

$$I_2(v_1, v_2) = (Iv_1, -Iv_2).$$

Then

$$fI_2 = I_1 f \tag{1.47}$$

because

$$fI_2(v_1, v_2) = f(Iv_1, -Iv_2) = (v_2 - v_1, I(v_1 + v_2)), \tag{1.48}$$

$$I_1 f(v_1, v_2) = I_1(I(v_1 + v_2), v_1 - v_2) = (v_2 - v_1, I(v_1 + v_2)). \tag{1.49}$$

Comparing (1.49) and (1.48) we obtain (1.47). Thus the mapping f gives an isomorphism of the bundle $cr\xi$ (in the image) with the sum of two complex vector bundles: the first is ξ and the second summand is homeomorphic to ξ but with different complex structure defined by the mapping I',

$$I'(v) = -I(v).$$

This new complex structure on the bundle ξ is denoted by $\bar{\xi}$. The vector bundle $\bar{\xi}$ is called the *complex conjugate of the complex bundle* ξ. Note that the vector bundles ξ and $\bar{\xi}$ are isomorphic as real vector bundles, that is, the isomorphism is compatible with respect to the large structure group $\mathbf{GL}\,(2n, \mathbf{R})$ but not isomorphic with respect to the structure group $\mathbf{GL}\,(n, \mathbf{C})$.

Thus we have the formula

$$cr\xi = \xi \oplus \bar{\xi}.$$

The next proposition gives a description of the complex conjugate vector bundle in term of the transition functions.

Proposition 1 *Let the*

$$\varphi_{\beta\alpha} : U_{\alpha\beta} \longrightarrow \mathbf{GL}\,(n, \mathbf{C})$$

be transition functions of a complex vector bundle ξ. Then the complex conjugate vector bundle $\bar{\xi}$ is defined by the complex conjugate matrices $\bar{\varphi}_{\beta\alpha}$.

Proof.

Let V be a complex vector space of dimension n with the operator of multiplication by and imaginary unit I. Denote by V' the same space V with a new complex structure defined by the operator $I' = -I$. The complex basis $\{e_1, \ldots, e_n\}$ of the space V is simultaneously the complex basis of V'. But if the vector $v \in V$ has complex coordinates $(z_1, \ldots, z_n)$ then the same vector v in V' has coordinates $(\bar{z}_1, \ldots, \bar{z}_n)$.

Hence if $\{f_1, \ldots, f_n\}$ is another complex basis then the complex matrix of change from the first coordinate system to another is defined by expanding vectors $\{f_1, \ldots, f_n\}$ in terms of the basis $\{e_1, \ldots, e_n\}$.

For the space V these expansions are

$$f_k = \sum_j z_{jk} e_j, \qquad (1.50)$$

and in the space V' these expansions are

$$f_k = \sum_j \bar{z}_{jk} e_j. \qquad (1.51)$$

Relations (1.50) and (1.51) prove the proposition (1). ■

Proposition 2 *A complex vector bundle ξ has the form $\xi = c\eta$ if and only if there is a real linear mapping*

$$* : \xi \longrightarrow \xi \qquad (1.52)$$

such that

$$*^2 = 1, \quad *I = -I*, \qquad (1.53)$$

where I is the multiplication by the imaginary unit.

Proof.

First we prove the necessity. As a real vector bundle the bundle ξ has the form $\xi = \eta \oplus \eta$. Then define a mapping $* : \eta \oplus \eta \longrightarrow \eta \oplus \eta$ by formula

$$*(v_1, v_2) = (v_1, -v_2).$$

The mapping $*$ is called the *operation of complex conjugation* . If the vector v has the property $*(v) = v$ then v is called a *real vector*. Similarly, the section τ is called *real* if $*(\tau(x)) = \tau(x)$ for any $x \in B$.

A homomorphism $f : \xi \longrightarrow \xi$ is called *real* if $*f* = f$. If f is real and v is real then $f(v)$ is real as well.

The sufficiency we shall prove a little bit later.

1.4.2 Subbundles

Let $f : \xi_1 \longrightarrow \xi_2$ be a homomorphism of vector bundles with a common base B and assume that the fiberwise mappings $f_x : (\xi_1)_x \longrightarrow (\xi_2)_x$ have constant rank. Let

$$p_1 : E_1 \longrightarrow B$$
$$p_2 : E_2 \longrightarrow B$$

be the projections of the vector bundles ξ_1 and ξ_2. Put

$$
\begin{aligned}
E_0 &= \{y \in E_1 : f(y) = 0 \in p_2^{-1}(x),\ x = p_1(y)\} \\
E &= f(E_1).
\end{aligned}
$$

Theorem 7 *1. The mapping*

$$p_0 = p_1|E_0 : E_0 \longrightarrow B$$

is a locally trivial bundle which admits a unique vector bundle structure ξ_0 such that natural inclusion $E_0 \subset E_1$ is a homomorphism of vector bundles.

2. The mapping

$$p = p_2|E : E \longrightarrow B \tag{1.54}$$

is a locally trivial bundle which admits a unique vector bundle structure ξ such that the inclusion $E \subset E_2$ and the mapping $f : E_1 \longrightarrow E$ are homomorphisms of vector bundles.

3. There exist isomorphisms

$$
\begin{aligned}
\varphi : \xi_1 &\longrightarrow \xi_0 \oplus \xi, \\
\psi : \xi_2 &\longrightarrow \xi \oplus \eta,
\end{aligned}
$$

such that composition

$$\psi \circ f \circ \varphi^{-1} : \xi_0 \oplus \xi \longrightarrow \xi \oplus \eta$$

has the matrix form

$$\psi \circ f \circ \varphi^{-1} = \begin{pmatrix} 0 & \mathbf{Id} \\ 0 & 0 \end{pmatrix}.$$

The bundle ξ_0 is called the *kernel* of the mapping f and denoted **Ker** f, the bundle ξ is called the *image* of the mapping f and denoted **Im** f. So we have

$$\dim \xi_1 = \dim \mathbf{Ker}\, f + \dim \mathbf{Im}\, f.$$

Proof.

Assume that vector bundles ξ_1 and ξ_2 are equipped with scalar products. Firstly, let us show that the statement 2 holds, that is, the mapping (1.54) gives a vector bundle. It is sufficient to prove that for any point $x_0 \in B$ there is a chart $U \ni x_0$ and a system of continuous sections $\sigma_1, \ldots, \sigma_k : U \longrightarrow E$ which form a basis in each subspace $f_x(p_1^{-1}(x)) \subset p_2^{-1}(x)$. For this fix a basis $\{e_1, \ldots, e_k\}$ in the subspace $f_{x_0}(p_1^{-1}(x_0))$. The mapping

$$f_{x_0} : p_1^{-1}(x_0) \longrightarrow f_{x_0}(p_1^{-1}(x_0))$$

is an epimorphism and hence we may choose vectors $g_1, \ldots, g_k \in p_1^{-1}(x_0)$ such that

$$f_{x_0}(g_j) = e_j, \; 1 \le j \le k.$$

Since ξ_1 is a locally trivial bundle for each x_0 there is a chart $U \ni x_0$ such that over U the bundle ξ_1 is isomorphic to a Cartesian product $U \times p_1^{-1}(x_0)$. Therefore there are continuous sections

$$\tau_1, \ldots, \tau_k : U \longrightarrow E_1$$

such that

$$\tau_j(x_0) = g_j, \; 1 \le j \le k.$$

Consider the sections

$$\sigma_j = f(\tau_j), \; 1 \le j \le k.$$

Then

$$\sigma_j(x_0) = f(\tau_j(x_0)) = f(g_j) = e_j, \; 1 \le j \le k.$$

The sections σ_j are continuous and hence there is a smaller neighborhood $U' \ni x_0$ such that for any point $x \in U'$ the system $\{\sigma_1(x), \ldots, \sigma_k(x)\}$ forms a basis of the subspace $f_x(p_1^{-1}(x))$. Thus we have shown that E is the total space of a vector bundle and the inclusion $E \subset E_2$ is a homomorphism. Uniqueness of the vector bundle structure on the E is clear. Now we pass to the statement 1. Denote by $P : \xi_1 \longrightarrow \xi_1$ the fiberwise mapping which in each fiber $p_1^{-1}(x)$ is the orthonormal projection onto the kernel **Ker** $f_x \subset p_1^{-1}(x)$. It sufficient to check that P is continuous over each separate chart U. Notice that the kernel **Ker** f_x is the orthogonal complement of the subspace **Im** f_x^*. The rank of f_x^* equals the rank of f_x. Hence we can apply part 2 of Theorem 7 to the mapping f^*.

Thus the image **Im** f^* is a locally trivial vector bundle. Now let $\{\sigma_1, \ldots, \sigma_k\}$ be a system of sections,

$$\sigma_j : U \longrightarrow \mathbf{Im}\, f^* \subset E_1,$$

forming a basis of the subspace **Im** $f_x^* \subset p_1^{-1}(x)$ for any $x \in U$. Then applying the projection P to a vector $y \in p_1^{-1}(x)$ gives a decomposition of the vector y into a linear combination

$$y = P(y) + \sum_{j=1}^{k} \lambda_j \sigma_j(x)$$

satisfying

$$\langle P(y), \sigma_j(x) \rangle = 0,\ 1 \le j \le k$$

or

$$\langle y - \sum_{j=1}^{k} \lambda_j \sigma_j(x), \sigma_l(x) \rangle = 0,\ 1 \le l \le k.$$

The coefficients λ_j satisfy the following system of linear equations

$$\sum_{j=1}^{k} \lambda_j \langle \sigma_j(x), \sigma_l(x) \rangle = \langle y, \sigma_l(x) \rangle, 1 \le l \le k.$$

The matrix $\|\langle \sigma_j(x), \sigma_l(x) \rangle\|$ depends continuously on x. Hence the inverse matrix also depends continuously and, as a consequence, the numbers λ_j depend continuously on x. It follows that the projection P is continuous. The rank of P does not depend on x and so we can apply statement 2. Thus statement 1 is proved. The last statement is implied by the following decompositions

$$\begin{aligned}
\xi_1 &= \mathbf{Ker}\, f \oplus \mathbf{Im}\, f^*, \\
\xi_2 &= \mathbf{Ker}\, f^* \oplus \mathbf{Im}\, f.
\end{aligned} \tag{1.55}$$

Now we can complete the proof of Proposition 2. Consider a vector bundle ξ with mapping (1.52) satisfying the condition (1.53). Consider a mapping

$$P = \frac{1}{2}(1 + *) : \xi \longrightarrow \xi. \tag{1.56}$$

The mapping P is a projection in each fiber and has a constant rank. Applying Theorem 7 we get

$$\xi = \mathbf{Im}\, P \oplus \mathbf{Ker}\, P = \xi_0 \oplus \xi_1$$

and
$$I(\xi_0) = \xi_1, \ I(\xi_1) = \xi_0.$$
Hence
$$\xi = c\xi_0.$$
The proof of Proposition 2 is now completed.

Theorem 8 *Let ξ be a vector bundle over a compact base B. Then there is a vector bundle η over B such that*
$$\xi \oplus \eta = \bar{N} = \text{a trivial bundle.}$$

Proof.

Let us use Theorem 7. It is sufficient to construct a homomorphism
$$f : \xi \longrightarrow \bar{N},$$
where the rank of f equals $\dim \xi$ in each fiber. Notice that if ξ is trivial then such an f exists. Hence for any chart U_α there is a homomorphism
$$f_\alpha : \xi | U_\alpha \longrightarrow \bar{N}_\alpha, \ \text{rank } f_\alpha = \dim \xi.$$
Let $\{\varphi_\alpha\}$ be a partition of unity subordinate to the atlas $\{U_\alpha\}$. Then each mapping $\varphi_\alpha f_\alpha$ can be extended by the zero trivial mapping to a mapping
$$g_\alpha : \xi \longrightarrow \bar{N}_\alpha, \ g_\alpha | U_\alpha = \varphi_\alpha f_\alpha.$$
The mapping g_α has the following property: if $\varphi_\alpha(x) \neq 0$ then rank $g_\alpha|_x = \dim \xi$. Let
$$\begin{aligned} g : \xi \quad &\longrightarrow \quad \oplus_\alpha \bar{N}_\alpha = \bar{N}, \\ g \quad &= \quad \oplus_\alpha g_\alpha \ \text{that is} \\ g(y) \quad &= \quad (g_1(y), \ldots, g_\alpha(y), \ldots). \end{aligned}$$
It is clear that the rank of g at each point satisfies the relation
$$\text{rank } g_\alpha \leq \text{rank } g \leq \dim \xi.$$
Further, for each point $x \in B$ there is such an α that $\varphi_\alpha(x) \neq 0$. Hence,
$$\text{rank } g \equiv \dim \xi. \tag{1.57}$$
Therefore we can apply Theorem 7. By (1.57) we get
$$\textbf{Ker } g = 0.$$
Hence the bundle ξ is isomorphic to **Im** g and $\bar{N} = \textbf{Im } g \oplus \eta$. ∎

Theorem 9 *Let ξ_1, ξ_2 be two vector bundles over a base B and let $B_0 \subset B$ a closed subspace. Let*

$$f_0 : \xi_1|B_0 \longrightarrow \xi_2|B_0$$

be a homomorphism of the restrictions of the bundles to the subspace B_0. Then the mapping f_0 can be extended to a homomorphism

$$f : \xi_1 \longrightarrow \xi_2, \quad f|B_0 = f_0.$$

Proof.

In the case where the bundles are trivial the problem is reduced to constructing extensions of matrix valued functions. This problem is solved using Urysohn's lemma concerning the extension of continuous functions. For the general case, we apply Theorem 8. Let $\xi_1 \oplus \eta_1$ and $\xi_2 \oplus \eta_2$ be trivial bundles, let $P : \xi_2 \oplus \eta_2 \longrightarrow \xi_2$ be a natural projection, and let $Q : \xi_1 \longrightarrow \xi_1 \oplus \eta_1$ be the natural inclusion. Finally, let

$$h_0 = f_0 \oplus 0 : (\xi_1 \oplus \eta_1)|B_0 \longrightarrow (\xi_2 \oplus \eta_2)|B_0$$

be the direct sum of f_0 and the trivial mapping. Then there is an extension of h_0 to a homomorphism

$$h : \xi_1 \oplus \eta_1 \longrightarrow \xi_2 \oplus \eta_2.$$

Let

$$f = PhQ : \xi_1 \longrightarrow \xi_2. \tag{1.58}$$

It is clear that $f|B_0 = f_0$. ∎

Remark

Theorem 9 is true not only for compact bases but for more general spaces (for example, paracompact spaces). The proof then becomes a little bit more complicated.

Exercises

1. Using the notation of Theorem 9 and assuming that $f_0 : \xi_1|B_0 \longrightarrow \xi_2|B_0$ is a fiberwise monomorphism, prove that f_0 can be extended to a fiberwise monomorphism $f : \xi_1|U \longrightarrow \xi_2|U$ in some neighborhood $U \supset B_0$.

2. Under the conditions of previous exercise prove there is a trivial bundle $\bar{N}$ and a fiberwise monomorphism

$$g : \xi_1 \longrightarrow \xi_2 \oplus \bar{N},$$

extending f_0, that is,

$$g|B_0 = (f_0, 0).$$

1.5 VECTOR BUNDLES RELATED TO MANIFOLDS

The most natural vector bundles arise from the theory of smooth manifolds. Recall that by an *n–dimensional manifold* one means a metrizable space X such that for each point $x \in X$ there is an open neighborhood $U \ni x$ homeomorphic to an open subset V of n–dimensional linear space $\mathbf{R}^n$. A homeomorphism

$$\varphi : U \longrightarrow V \subset \mathbf{R}^n$$

is called a *coordinate homeomorphism* . The coordinate functions on the linear space $\mathbf{R}^n$ pulled back to points of the neighborhood U, that is, the compositions

$$x^j = x^j \circ \varphi : U \longrightarrow \mathbf{R}^1$$

are called *coordinate functions on the manifold X in the neighborhood U* This system of functions $\{x^1, \ldots, x^n\}$ defined on the neighborhood U is called a *local system of coordinates of the manifold X*. The open set U equipped with the local system of coordinates $\{x^1, \ldots, x^n\}$ is called a *chart* . The system of charts $\{U_\alpha, \{x_\alpha^1, \ldots, x_\alpha^n\}\}$ is called an *atlas* if $\{U_\alpha\}$ covers the manifold X, that is,

$$X = \cup_\alpha U_\alpha.$$

So each n–dimensional manifold has an atlas. If a point $x \in X$ belongs to two charts,

$$x \in U_\alpha \cap U_\beta,$$

then in a neighborhood of x there are two local systems of coordinates. In this case, the local coordinates x_α^j can expressed as functions of values of the local coordinates $\{x_\beta^1, \ldots, x_\beta^n\}$, that is, there are functions $f_{\alpha\beta}^k$ such that

$$x_\alpha^k = \varphi_{\alpha\beta}^k(x_\beta^1, \ldots, x_\beta^n). \tag{1.59}$$

The system of functions (1.59) are called a *change of coordinates* or *transition functions from one local coordinate system to another*. For brevity (1.59) will be written as

$$x_\alpha^k = x_\alpha^k(x_\beta^1, \ldots, x_\beta^n).$$

If an atlas $\{U_\alpha, \{x_\alpha^1, \ldots, x_\alpha^n\}\}$ is taken such that all the transition functions are differentiable functions of the class C^k, $1 \le k \le \infty$ then one says X has the *structure of differentiable manifold of the class C^k*. If all the transition functions are analytic functions then one says that X has the *structure of an analytic manifold*. In the case

$$
\begin{aligned}
n &= 2m, \\
u_\alpha^k &= x_\alpha^k, \ 1 \le k \le m, \\
v_\alpha^k &= x_\alpha^{m+k}, \ 1 \le k \le m, \\
z_\alpha^k &= u_\alpha^k + iv_\alpha^k, \ 1 \le k \le m,
\end{aligned}
$$

and the functions

$$z_\alpha^k = \varphi_{\alpha\beta}^k(z_\beta^1, \ldots, z_\beta^m) + i\varphi_{\alpha\beta}^{m+k}(z_\beta^1, \ldots, z_\beta^m)$$

are complex analytic functions in their domain of definition then X has the *structure of a complex analytic manifold*. Usually we shall consider infinitely smooth manifolds, that is, differentiable manifolds of the class C^∞. A mapping $f : X \longrightarrow Y$ of differentiable manifolds is called *differentiable of class C^k* if in any neighborhood of the point $x \in X$ the functions which express the coordinates of the image $f(x)$ in terms of coordinates of the point x are differentiable functions of the class C^k. It is clear that the class of differentiability k makes sense provided that k does not exceed the differentiability classes of the manifolds X and Y. Similarly, one can define analytic and complex analytic mappings.

Let X be a n–dimensional manifold, $\{U_\alpha, \{x_\alpha^1, \ldots, x_\alpha^n\}\}$ be an atlas. Fix a point $x_0 \in X$. A *tangent vector* ξ to the manifold X at the point x_0 is a system of numbers $(\xi_\alpha^1, \ldots, \xi_\alpha^n)$ satisfying the relations:

$$\xi_\alpha^k = \sum_{j=1}^n \xi_\beta^j \frac{\partial x_\alpha^k}{\partial x_\beta^j}(x_0). \tag{1.60}$$

The numbers $(\xi_\alpha^1, \ldots, \xi_\alpha^n)$ are called the *coordinates* or *components of the vector ξ with respect to the chart* $\{U_\alpha, \{x_\alpha^1, \ldots, x_\alpha^n\}\}$. The formula (1.60) gives the transformation law of the components of the tangent vector ξ under the transition from one chart to another. In differential geometry such a law is called a *tensor law of transformation of components of a tensor of the valency* $(1, 0)$. So in terms of differential geometry, a tangent vector is a tensor of the valency $(1,0)$. Consider a smooth curve γ passing through a point x_0, that is, a smooth mapping of the interval

$$\gamma : (-1,1) \longrightarrow X, \ \gamma(0) = x_0.$$

In terms of a local coordinate system $\{x_\alpha^1, \ldots, x_\alpha^n\}$ the curve γ is determined by a family of smooth functions

$$x_\alpha^k(t) = x_\alpha^k(\gamma(t)), \ t \in (-1,1).$$

Let

$$\xi_\alpha^k = \frac{\partial}{\partial t}(x_\alpha^k(t))|_{t=0}. \tag{1.61}$$

Clearly the numbers (1.61) satisfy a tensor law (1.60), that is, they define a tangent vector ξ at the point x_0 to manifold X. This vector is called the *tangent vector to the curve* γ and is denoted by $\frac{d\gamma}{dt}(0)$, that is,

$$\xi = \frac{d\gamma}{dt}(0).$$

The family of all tangent vectors to manifold X is a denoted by TX. The set TX is endowed with a natural topology. Namely, a neighborhood V of the vector ξ_0 at the point x_0 contains all vectors η in points x such that $x \in U_\alpha$ for some chart U_α and for some ε,

$$\rho(x, x_0) < \varepsilon,$$

$$\sum_{k=1}^{n} (\xi_{0\alpha}^k - \eta_\alpha^k)^2 < \varepsilon.$$

The verification that the system of the neighborhoods V forms a base of a topology is left to the reader.

Let

$$\pi : TX \longrightarrow X \tag{1.62}$$

be the mapping which to any vector ξ associates its point x of tangency. Clearly, the mapping π is continuous. Moreover, the mapping (1.62) defines a locally trivial vector bundle with the base X, total space TX and fiber isomorphic to the linear space $\mathbf{R}^n$. If U_α is a chart on the manifold X, we define a homeomorphism

$$\varphi_\alpha : U_\alpha \times \mathbf{R}^n \longrightarrow \pi^{-1}(U_\alpha)$$

which to the system $(x_0, \xi^1, \ldots, \xi^n)$ associates the tangent vector ξ whose components are defined by the formula

$$\xi_\beta^k = \sum_{j=1}^{n} \xi^j \frac{\partial x_\beta^k}{\partial x_\alpha^j}(x_0). \tag{1.63}$$

It is easy to check that the definition (1.63) gives the components of a vector ξ, that is, they satisfy the tensor law for the transformation of components of a tangent vector (1.60). The inverse mapping is defined by the following formula:

$$\varphi_\alpha^{-1}(\xi) = (\pi(\xi), \xi_\alpha^1, \ldots, \xi_\alpha^n),$$

where ξ_α^k are components of the vector ξ. Therefore the transition functions $\varphi_{\beta\alpha} = \varphi_\beta^{-1}\varphi_\alpha$ are determined by the formula

$$\varphi_{\beta\alpha}(x_0, \xi^1, \ldots, \xi^n) = \left(x_0, \sum \xi^j \frac{\partial x_\beta^1}{\partial x_\alpha^j}(x_0), \ldots, \sum \xi^j \frac{\partial x_\beta^n}{\partial x_\alpha^j}(x_0) \right). \qquad (1.64)$$

Formula (1.64) shows that the transition functions are fiberwise linear mappings. Hence the mapping π defines a vector bundle. The vector bundle $\pi : TX \longrightarrow X$ is called the *tangent bundle of the manifold* X. The fiber $T_x X$ is called the *tangent space at the point* x *to manifold* X.

The terminology described above is justified by the following. Let $f : X \longrightarrow \mathbf{R}^N$ be an inclusion of the manifold X in the Euclidean space $\mathbf{R}^N$. In a local coordinate system $(x_\alpha^1, \ldots, x_\alpha^n)$ in a neighborhood of the point $x_{0\in X}$, the inclusion f is determined as a vector valued function of the variables $(x_\alpha^1, \ldots, x_\alpha^n)$:

$$f(x) = f(x_\alpha^1, \ldots, x_\alpha^n). \qquad (1.65)$$

If we expand the function (1.65) by a Taylor expansion at the point $x_0 = (x_{0\alpha}^1, \ldots, x_{0\alpha}^n)$:

$$f(x_\alpha^1, \ldots, x_\alpha^n) = f(x_{0\alpha}^1, \ldots, x_{0\alpha}^n) +$$
$$+ \sum_{j=1}^n \frac{\partial f}{\partial x_\alpha^j}(x_{0\alpha}^1, \ldots, x_{0\alpha}^n)\Delta x_\alpha^j + o(\Delta x_\alpha^k).$$

Ignoring the remainder term $o(\Delta x_\alpha^k)$ we obtain a function g which is close to f :

$$g(x_\alpha^1, \ldots, x_\alpha^n) = f(x_{0\alpha}^1, \ldots, x_{0\alpha}^n) +$$
$$+ \sum_{j=1}^n \frac{\partial f}{\partial x_\alpha^j}(x_{0\alpha}^1, \ldots, x_{0\alpha}^n)\Delta x_\alpha^j.$$

Then if the vectors

$$\left\{ \frac{\partial f}{\partial x_\alpha^j}(x_{0\alpha}^1, \ldots, x_{0\alpha}^n)\Delta x_\alpha^j \right\}, \quad 1 \leq k \leq n$$

are linearly independent the function g defines a linear n–dimensional subspace in $\mathbf{R}^n$. It is natural to call this space the *tangent space to the manifold* X. Any vector ξ which lies in the tangent space to manifold X (having the initial point at x_0) can be uniquely decomposed into a linear combination of the basis vectors:

$$\xi = \sum_{j=1}^{n} \frac{\partial f}{\partial x_\alpha^j} \xi_\alpha^j. \tag{1.66}$$

The coordinate s $\{\xi_\alpha^k\}$ under a change of coordinate system change with respect to the law (1.60), that is, with respect to the tensor law. Thus the abstract definition of tangent vector as a system of components $\{\xi_\alpha^k\}$ determines by the formula (1.66) a tangent vector to the submanifold X in $\mathbf{R}^N$. Let

$$f : X \longrightarrow Y$$

be a differentiable mapping of manifolds. Let us construct the corresponding homomorphism of tangent bundles,

$$Df : TX \longrightarrow TY.$$

Let $\xi \in TX$ be a tangent vector at the point x_0 and let γ be a smooth curve which goes through the point x_0,

$$\gamma(0) = x_0,$$

and which has tangent vector ξ, that is,

$$\xi = \frac{d\gamma}{dt}(0).$$

Then the curve $f(\gamma(t))$ in the manifold Y goes through the point $y_0 = f(x_0)$. Put

$$Df(\xi) = \frac{d(f(\gamma))}{dt}(0). \tag{1.67}$$

This formula (1.67) defines a mapping of tangent spaces. It remains to prove that this mapping is fiberwise linear. For this, it is sufficient to describe the mapping Df in terms of coordinates of the spaces $T_{x_0}X$ and $T_{y_0}Y$. Let $\{x_\alpha^1, \ldots, x_\alpha^n\}$ and $\{y_\beta^1, \ldots, y_\beta^m\}$ be local systems of coordinates in neighborhoods of points the x_0 and y_0, respectively . Then the mapping f is defined as a family of functions

$$y_\beta^j = y_\beta^j(x_\alpha^1, \ldots, x_\alpha^n).$$

If

$$x_\alpha^k = x_\alpha^k(t)$$

are the functions defining the curve $\gamma(t)$ then

$$\xi_\alpha^k = \frac{dx_\alpha^k}{dt}(0).$$

Hence the curve $f(\gamma(t))$ is defined by the functions

$$y_\beta^j = y_\beta^j(x_\alpha^1(t), \ldots, x_\alpha^n(t))$$

and the vector $Df(\xi)$ is defined by the components

$$\eta^j = \frac{dy_\beta^j}{dt}(0) = \sum_{k=1}^n \frac{\partial y_\beta^j}{\partial x_\alpha^k}(x_0)\frac{dx_\alpha^k}{dt}(0) =$$

$$\sum_{k=1}^n \frac{\partial y_\beta^j}{\partial x_\alpha^k}(x_0)\xi^k. \tag{1.68}$$

Formula (1.68) shows firstly that the mapping Df is well-defined since the definition does not depend on the choice of curve γ but only on the tangent vector at the point x_0. Secondly, the mapping Df is fiberwise linear. The mapping Df is called the *differential of the mapping f*.

Examples

1. Let us show that the definition of differential Df of the mapping f is a generalization of the notion of the classical differential of function. A differentiable function of one variable may be considered as a mapping of the space $\mathbf{R}^1$ into itself:

$$f : \mathbf{R}^1 \longrightarrow \mathbf{R}^1.$$

The tangent bundle of the manifold $\mathbf{R}^1$ is isomorphic to the Cartesian product $\mathbf{R}^1 \times \mathbf{R}^1 = \mathbf{R}^2$. Hence the differential

$$Df : \mathbf{R}^1 \times \mathbf{R}^1 \longrightarrow \mathbf{R}^1 \times \mathbf{R}^1$$

in the coordinates (x, ξ) is defined by the formula

$$Df(x, \xi) = (x, f'(x)\xi).$$

The classical differential has the form

$$df = f'(x)dx.$$

So

$$Df(x, dx) = (x, df).$$

Exercises

Show that differential Df has the following properties:

1. $D(f \circ g) = (Df) \circ (Dg)$,

2. $D(\mathbf{Id}) = \mathbf{Id}$,

3. if f is a diffeomorphism the Df is isomorphism of bundles,

4. if f is immersion then Df is fiberwise a monomorphism.

Consider a smooth manifold Y and a submanifold $X \subset Y$. The inclusion

$$j : X \subset Y$$

is a smooth mapping of manifolds such that the differential

$$Dj : TX \longrightarrow TY$$

is a fiberwise monomorphism. Then over the manifold X there are two vector bundles: the first is $j^*(TY)$, the restriction of the tangent bundle of the manifold Y to the submanifold X, the second is its subbundle TX.

According to Theorem 7, the bundle $j^*(TY)$ splits into a direct sum of two summands:
$$j^*(TY) = TX \oplus \eta.$$

The complement η is called the *normal bundle to the submanifold X of the manifold Y* [1]. Each fiber of the bundle η over the point x_0 consists of those tangent vectors to the manifold Y which are orthogonal to the tangent space $T_{x_0}(X)$. The normal bundle will be denoted by $\nu(X \subset Y)$ or more briefly by $\nu(X)$.

It is clear that the notion of a normal bundle can be defined not only for submanifolds but for any immersion $j : X \longrightarrow Y$ of the manifold X into the manifold Y. It is known that any compact manifold X has an inclusion in a Euclidean space $\mathbf{R}^N$ for some sufficiently large number N. Let $j : X \longrightarrow \mathbf{R}^N$ be such an inclusion. Then

$$j^*(T\mathbf{R}^N) = TX \oplus \nu(X \subset \mathbf{R}^N).$$

---

[1] By construction the bundle η depends on the choice of metric on TY. However, different metrics yield isomorphic complementary summands

The bundle $T\mathbf{R}^N$ is trivial and so

$$TX \oplus \nu(X) = \bar{N}. \tag{1.69}$$

In this case the bundle $\nu(X)$ is called *the normal bundle for manifold* X (irrespective of the inclusion). Note, the normal bundle $\nu(X)$ of the manifold X is not uniquely defined. It depends on inclusion into the space $\mathbf{R}^N$ and on the dimension N. But the equation (1.69) shows that the bundle is not very far from being unique.

Let $\nu_1(X)$ be another such normal bundle, that is,

$$TX \oplus \nu_1(X) = \bar{N}_1.$$

Then

$$\nu(X) \oplus T(X) \oplus \nu_1(X) = \nu(X) \oplus \bar{N}_1 = \bar{N} \oplus \nu_1(X).$$

The last relation means that bundles $\nu(X)$ and $\nu_1(X)$ became isomorphic after the addition of trivial summands.

2. Let us study the tangent bundle of the one dimensional manifold $\mathbf{S}^1$, the circle. Define two charts on the $\mathbf{S}^1$:

$$U_1 = \{-\pi < \varphi < \pi\},$$
$$U_2 = \{0 < \varphi < 2\pi\},$$

where φ is angular parameter in the polar system of coordinates on the plane. On the U_1 take the function

$$x_1 = \varphi, \ -\pi < x_1 < \pi,$$

as coordinate, whereas on the U_2 take the function

$$x_2 = \varphi, \ 0 < x_2 < 2\pi.$$

The intersection $U_1 \cap U_2$ consists of the two connected components

$$V_1 = \{0 < \varphi < \pi\},$$

$$V_2 = \{\pi < \varphi < 2\pi\}.$$

Then the transition function has the form

$$x_1 = x_1(x_2) = \begin{cases} x_2, & 0 < x_2 < \pi, \\ x_2 - 2\pi, & \pi < x_2 < 2\pi. \end{cases}$$

Then by (1.64), the transition function φ_{12} for the tangent bundle has the form

$$\varphi_{12}(x,\xi) = \xi \frac{\partial x_1}{\partial x_2} = \xi.$$

This means that the transition function is the identity. Hence the tangent bundle TS^1 is isomorphic to Cartesian product

$$TS^1 = S^1 \times R^1,$$

in other words, it is the trivial one dimensional bundle.

3. Consider the two-dimensional sphere S^2. It is convenient to consider it as the extended complex plane

$$S^2 = C^1 \cup \{\infty\}.$$

We define two charts on the S^2

$$\begin{aligned} U_1 &= C^1 \\ U_2 &= (C^1 \backslash \{0\}) \cup \{\infty\}. \end{aligned}$$

Define the complex coordinate $z_1 = z$ on the chart U_1 and $z_2 = \frac{1}{z}$ on the chart U_2 extended by zero at the infinity ∞. Then the transition function on the intersection $U_1 \cap U_2$ has the form

$$z_1 \equiv \frac{1}{z_2}$$

and the tangent bundle has the corresponding transition function of the form

$$\varphi_{12}(z,\xi) = \xi \frac{\partial z_1}{\partial z_2} = -\xi \frac{1}{z_2^2} = -\xi z^2.$$

The real form of the matrix φ_{12} is given by

$$\varphi_{12}(x,y) = \left\| \begin{array}{cc} -\Re z^2 & -\Im z_2 \\ \Im x_2 & -\Re z_2 \end{array} \right\| = \left\| \begin{array}{cc} y_2 - x_2 & -2xy \\ 2xy & y_2 - x_2 \end{array} \right\|,$$

where $z = x + iy$. In polar coordinates $z = \rho e^{i\alpha}$ this becomes

$$\varphi_{12}(\rho,\alpha) = \rho^2 \left\| \begin{array}{cc} \cos 2\alpha & -\sin 2\alpha \\ \sin 2\alpha & \cos 2\alpha \end{array} \right\|.$$

Let us show that the tangent bundle TS^2 is not isomorphic to a trivial bundle. If the bundle TS^2 were trivial then there would be matrix valued functions

$$\begin{aligned} h_1 : U_1 &\longrightarrow GL(2, R) \\ h_2 : U_2 &\longrightarrow GL(2, R), \end{aligned} \tag{1.70}$$

such that
$$\varphi_{12}(\rho, \alpha) = h_1(\rho, \alpha)h_2^{-1}(\rho, \alpha).$$

The charts U_1, U_2 are contractible and so the functions h_1, h_2 are homotopic to constant mappings.

Hence the transition function $\varphi_{12}(\rho, \alpha)$ must be homotopic to a constant function. On the other hand, for fixed ρ the function φ_{12} defines a mapping of the circle $\mathbf{S}^1$ with the parameter α into the group $\mathbf{SO}(2) = \mathbf{S}^1$ and this mapping has the degree 2. Therefore this mapping cannot be homotopic to a constant mapping.

4.   Consider a vector bundle $p : E \longrightarrow X$ where the base X is a smooth manifold. Assume that the transition functions

$$\varphi_{\alpha\beta} : U_\alpha \cap U_\beta \longrightarrow \mathbf{GL}\,(n, \mathbf{R})$$

are smooth mappings. Then the total space E is a smooth manifold and also

$$\dim E = \dim X + n.$$

For if $\{U_\alpha\}$ is an atlas of charts for the manifold X then an atlas of the manifold E can be defined by
$$V_\alpha = p^{-1}(U_\alpha) = U_\alpha \times \mathbf{R}^n.$$

The local coordinates on V_α can be defined as the family of local coordinates on the chart U_α with Cartesian coordinates on the fiber. The smoothness of the functions $\varphi_{\alpha\beta}$ implies smoothness of the change of coordinates.

There is the natural question whether for any vector bundle over smooth manifold X there exists an atlas on the total space with smooth the transition functions $\varphi_{\alpha\beta}$. The answer lies in the following theorem.

Theorem 10 *Let $p : E \longrightarrow X$ be an n–dimensional vector bundle and X a compact smooth manifold. Then there exists an atlas $\{U_\alpha\}$ on X and coordinate homeomorphisms*

$$\varphi_\alpha : U_\alpha \times \mathbf{R}^n \longrightarrow p^{-1}(U_\alpha)$$

such that the transition functions

$$\varphi_{\alpha\beta} : U_\alpha \cap U_\beta \longrightarrow \mathbf{GL}\,(n, \mathbf{R})$$

are smooth.

Proof.

Consider a sufficiently fine atlas for the bundle p and coordinate homeomorphisms

$$\psi_\alpha : U_\alpha \times \mathbf{R}^n \longrightarrow p^{-1}(U_\alpha).$$

The transition functions

$$\psi_{\alpha\beta} : U_\alpha \cap U_\beta \longrightarrow \mathbf{GL}\,(n, \mathbf{R})$$

are in general only required to be continuous. The problem is to change the homeomorphisms ψ_α to new coordinate homeomorphisms φ_α such that the new transition functions $\varphi_{\alpha\beta}$ are smooth. In other words, one should find functions

$$h_\alpha : U_\alpha \longrightarrow \mathbf{GL}\,(n, \mathbf{R})$$

such that the compositions $h_\beta^{-1}(x)\psi_{\beta\alpha}(x)h_\alpha(x)$ are smooth. Let $\{U'_\alpha\}$ be a new atlas such that

$$U'_\alpha \subset \bar{U}'_\alpha \subset U_\alpha.$$

We construct the functions $h_\alpha(x)$ by induction on the index α, $1 \leq \alpha \leq N$. Without loss of generality, we can assume that all the functions $\psi_{\alpha\beta}(x)$ are uniformly bounded, that is,

$$\|\psi_{\alpha\beta}(x)\| \leq C.$$

Let

$$0 < \varepsilon < \frac{1}{C},\ \varepsilon < 1.$$

Choose smooth functions $\varphi_{\alpha\beta}(x)$, $x \in U_\alpha \cap U_\beta$ such that

$$\|\psi_{\alpha\beta}(x) - \varphi_{\alpha\beta}(x)\| < \varepsilon$$

and which form a cocycle, that is,

$$\varphi_{\alpha\beta}\varphi_{\beta\gamma}\varphi_{\gamma\alpha} \equiv 1,\ \varphi_{\alpha\beta}\varphi_{\beta\alpha} \equiv 1.$$

Lemma 2 *Let $f(x)$ be a continuous function defined in a domain U of Euclidean space, let K be a compact subset such that $K \subset U$. Then there is a neighborhood $V \supset K$ such that for each $\varepsilon > 0$ there is a smooth function $g(x)$ defined on V such that*

$$|g(x) - f(x)| < \varepsilon,\ x \in V.$$

Moreover, if the function $f(x)$ is smooth in a neighborhood of a compact subset K' then the function g can be chosen with additional property that

$$g(x) \equiv f(x)$$

in a neighborhood of K'.

Let us apply the lemma (2) for construction of functions $\varphi_{\alpha\beta}$. By the lemma there exists a smooth function φ_{12} defined in a neighborhood of the set $\bar{U}'_1 \cap \bar{U}'_2$ such that

$$\|\varphi_{12}(x) - \psi_{12}(x)\| < \varepsilon.$$

Assume that we have chosen smooth functions $\varphi_{\alpha\beta}(x)$ defined in neighborhoods of the subsets $\bar{U}'_\alpha \cap \bar{U}'_\beta$ for all indices $\alpha < \beta \le \beta_0$ such that

$$\|\varphi_{\alpha\beta}(x) - \psi_{\alpha\beta}(x)\| < \varepsilon$$

and

$$\varphi_{\alpha\beta}(x)\varphi_{\beta\gamma}(x)\varphi_{\gamma\alpha}(x) \equiv 1, \alpha, \beta, \gamma \le \beta_0.$$

The function $\varphi_{1,\beta_0+1}(x)$ can be constructed in the same way as φ_{12} using the lemma (2).

Assume that functions $\varphi_{\alpha,\beta_0+1}(x)$ are constructed for all $\alpha \le \alpha_0 \le \beta_0$ and they satisfy the condition of cocycle:

$$\varphi_{\alpha,\beta_0+1}(x) \equiv \varphi_{\alpha,\alpha'}(x)\varphi_{\alpha',\beta_0+1}(x), \ \alpha, \alpha' \le a_0.$$

Then the required function $\varphi_{\alpha_0+1,\beta_0+1}(x)$ can be defined in a neighborhood of the subset $\bar{U}'_\gamma \cap \bar{U}'_{\alpha_0+1} \cap \bar{U}'_{\beta_0+1}$ for any $\gamma \le \alpha_0$ by formula

$$\varphi_{\alpha_0+1,\beta_0+1}(x) = \varphi^{-1}_{\gamma,\alpha_0+1}(x)\varphi_{\gamma,\beta_0+1}(x).$$

Hence the required function $\varphi_{\alpha_0+1,\beta_0+1}(x)$ is defined in a neighborhood of the union

$$V = \cup_{\gamma \le \alpha_0}(\bar{U}'_\gamma \cap \bar{U}'_{\alpha_0+1} \cap \bar{U}'_{\beta_0+1})$$

and satisfies the condition

$$\|\psi_{\alpha_0+1,\beta_0+1}(x) - \varphi_{\alpha_0+1,\beta_0+1}(x)\| < 2C\varepsilon. \tag{1.71}$$

By the lemma (2) the function can be extended $\varphi_{\alpha_0+1,\beta_0+1}(x)$ from the closure of a neighborhood of the set V to a neighborhood of the set $\bar{U}'_{\alpha_0+1} \cap \bar{U}'_{\beta_0+1}$ satisfying the same condition (1.71).

By induction, a family of functions $\varphi_{\alpha\beta}$ with property

$$\|\psi_{\alpha,\beta}(x) - \varphi_{\alpha,\beta}(x)\| < (2C)^{N^2}\varepsilon = \varepsilon'.$$

can be constructed. Now we pass to the construction of functions h_α satisfying the conditions

$$\varphi_{\alpha\beta}(x) = h_\alpha(x)\psi_{\alpha\beta}(x)h_\beta^{-1}(x)$$

or

$$h_\alpha(x) = \varphi_{\alpha\beta}(x)h_\beta(x)\psi_{\beta\alpha}(x). \tag{1.72}$$

We construct the functions h_α by induction. Put

$$h_1(x) \equiv 1.$$

Assume that the functions $h_\alpha(x), 1 \le \alpha \le \alpha_0$ have already been constructed satisfying the conditions (1.72) and

$$\|1 - h_\alpha(x)\| < \varepsilon, \ \alpha < \alpha_0.$$

Then the function $h_{\alpha_0+1}(x)$ is defined by the formula (1.72)in a neighborhood of the set

$$V = \cup_{\alpha < \alpha_0} \bar{U}'_{\alpha_0+1} \cap \bar{U}'_\alpha.$$

On the set V, the inequality

$$\|1 - h_{\alpha_0+1}(x)\| = \|1 - \varphi_{\alpha_0+1,\beta}(x)h_\beta(x)\psi_{\beta,\alpha_0+1}(x)\| \le$$
$$\le \ C^2\|h_\beta - \varphi_{\beta,\alpha_0+1}\psi_{\beta,\alpha_0+1}^{-1}\| \le 2C^2\varepsilon \tag{1.73}$$

is satisfied. If the number ε is sufficient small then the function $h_{\alpha_0+1}(x)$ can be extended from a neighborhood V on the whole chart U_{α_0+1} satisfying the same condition (1.73). ∎

5. Let us describe the structure of the tangent and normal bundles of a smooth manifold.

Theorem 11 *Let $j : X \subset Y$ be a smooth submanifold X of a manifold Y. Then there exists a neighborhood $V \supset X$ which is diffeomorphic to the total space of the normal bundle $\nu(X \subset Y)$.*

Proof.

Fix a Riemannian metric on the manifold Y (which exists by Theorem 5 and Remark 2 from the section 3). We construct a mapping

$$f : \nu(X) \longrightarrow Y.$$

Consider a normal vector $\xi \in \nu(X)$ at the point $x \in X \subset Y$. Notice that the vector ξ is orthogonal to the subspace $T_x(X) \subset T_x(Y)$. Let $\gamma(t)$ be the geodesic curve such that

$$\gamma(0) = x, \ \frac{d\gamma}{dt}(0) = \xi.$$

Put $f(\xi) = \gamma(1)$. The mapping f has nondegenerate Jacobian for each point of the zero section of the bundle $\nu(X)$. Indeed, notice that

1. if $\xi = 0$, $\xi \in T_x(Y)$ then $f(\xi) = x$,

2. $f(\lambda\xi) = \gamma(\lambda)$.

Therefore the Jacobian matrix of the mapping f at a point of the zero section of $\nu(X)$ that maps the tangent spaces

$$Df : T_x(\nu(X)) = T_x(X) \oplus \nu_x(X) \longrightarrow T_x(Y) = T_x(X) \oplus \nu_x(X)$$

is the identity. By the implicit function theorem there is a neighborhood V of zero section o f the bundle $\nu(X)$ which is mapped by F diffeomorphically onto a neighborhood $f(V)$ of the submanifold X. Since there is a sufficiently small neighborhood V which is diffeomorphic to the total space of the bundle $\nu(X)$ the proof of theorem is finished.　∎

Exercise

Prove that the total space of the tangent bundle TX for the manifold X is diffeomorphic to a neighborhood V of the diagonal $\Delta \subset X \times X$.

6. There is a simple criterion describing when a smooth mapping of manifolds gives a locally trivial bundle.

Theorem 12 *Let*

$$f : X \longrightarrow Y$$

be a smooth mapping of compact manifolds such that the differential Df is epimorphism at each point $x \in X$. Then f is a locally trivial bundle with the fiber a smooth manifold.

Proof.

Without loss of generality one can consider a chart $U \subset Y$ diffeomorphic to $\mathbf{R}^n$ and part of the manifold X, namely, $f^{-1}(U)$. Then the mapping f gives a vector valued function

$$f : X \longrightarrow \mathbf{R}^n.$$

Assume firstly that $n = 1$. From the condition of the theorem we know that the gradient of the function f never vanishes. Consider the vector field grad f (with respect to some Riemannian metric on X). The integral curves $\gamma(x_0, t)$ are orthogonal to each hypersurface of the level of the function f. Choose a new Riemannian metric such that grad f is a unit vector field. Indeed, consider the new metric

$$(\xi, \eta)_1 = (\xi, \eta)(\text{grad } f, \text{grad } f).$$

Then

$$(\text{grad } f, \xi) = \xi(f) = (\text{grad }_1 f, \xi)_1 = (\text{grad }_1 f, \xi)(\text{grad }_1 f, \text{grad }_1 f).$$

Hence

$$\text{grad }_1 f = \frac{\text{grad } f}{(\text{grad } f, \text{grad } f)}.$$

Then

$$(\text{grad }_1 f, \text{grad }_1 f)_1 = (\text{grad }_1 f, \text{grad }_1 f)(\text{grad } f, \text{grad } f) =$$

$$= \frac{(\text{grad } f, \text{grad } f)}{(\text{grad } f, \text{grad } f)^2}(\text{grad } f, \text{grad } f) = 1.$$

Thus the integral curves

$$\frac{d}{dt} f(\gamma(t)) = (\text{grad } f, \text{grad } f) \equiv 1.$$

Hence the function $f(\gamma(t))$ is linear. This means that if

$$f(x_0) = f(x_1),$$

then

$$f(\gamma(x_0, t)) = f(\gamma(x_1, t)) = f(x_0) + t.$$

Put

$$g : Z \times \mathbf{R}^1 \longrightarrow X, \ g(x, t) = \gamma(x, t).$$

The mapping g is a fiberwise smooth homeomorphism. Hence the mapping

$$f : \longrightarrow \mathbf{R}^1$$

gives a locally trivial bundle. Further, the proof will follow by induction with respect to n. Consider a vector valued function

$$f(x) = \{f_1(x), \ldots, f_n(x)\}$$

which satisfies the condition of the theorem. Choose a Riemannian metric on the manifold X such that gradients

$$\operatorname{grad} f_1, \ldots, \operatorname{grad} f_n$$

are orthonormal. Such a metric exists. Indeed, consider firstly an arbitrary metric. The using the linear independence of the differentials $\{df_i\}$ we know that the gradients are also independent. Hence the matrix

$$a_{ij} = \langle \operatorname{grad} f_i(x), \operatorname{grad} f_j(x) \rangle$$

is nondegenerate in each point. Let $\|b_{ij}(x)\|$ be the matrix inverse to the matrix $\|a_{ij}(x)\|$, that is,

$$\sum_\alpha b_{i\alpha}(x) a_{j\alpha}(x) \equiv \delta_{ij}.$$

Put

$$\xi_k = \sum_i b_{kj}(x) \operatorname{grad} f_i(x).$$

Then

$$\xi_k(f_j) = \sum_i b_{ki}(x) \operatorname{grad} f_i(f_j) =$$

$$= \sum_i b_{ki}(x) \langle \operatorname{grad} f_i, \operatorname{grad} g_j \rangle = \sum_i b_{ki}(x) a_{ij}(x) \equiv \delta_{kj}.$$

Let U_α be a sufficiently small neighborhood of a point of the manifold X. The system of vector fields $\{\xi_1, \ldots, \xi_n\}$ can be supplemented by vector fields $\eta_{n+1}, \ldots, \eta_N$ to form a basis such that

$$\eta_k(f_i) \equiv 0.$$

Consider the new metric in the chart U_α given by

$$\langle \xi_i, \xi_j \rangle_\alpha \equiv \delta_{ij},$$
$$\langle \xi_k, \eta_j \rangle_\alpha \equiv 0.$$

Let φ_α be a partition of unity subordinate to the covering $\{U_\alpha\}$ and put

$$\langle \xi, \eta \rangle_0 = \sum_\alpha \varphi_\alpha(x) \langle \xi, \eta \rangle_\alpha.$$

Then

$$\langle \xi_i, \xi_j \rangle_0 \equiv \delta_{ij},$$
$$\langle \xi_k, \eta \rangle_0 \equiv 0$$

for any vector η for which $\eta(f_i) = 0$. Let grad $_0 f_i$ be the gradients of the functions f_i with respect to the metric $\langle \xi, \eta \rangle_0$. This means that

$$\langle \text{grad } f_i, \xi \rangle_0 = \xi(f_i)$$

for any vector ξ. In particular one has

$$\langle \text{grad } f_i, \xi_j \rangle_0 \equiv \delta_{ij},$$
$$\langle \text{grad } f_k, \eta \rangle_0 \equiv 0$$

for any vector η for which $\eta(f_i) = 0$. Similar relations hold for the vector field ξ_i. Therefore

$$\xi_i = \text{grad } f_i,$$

that is,

$$\langle \text{grad } f_i, \text{grad } f_j \rangle \delta_{ij},$$

the latter proves the existence of metric with the necessarily properties.

Let us pass now to the proof of the theorem. Consider the vector function

$$g(x) = \{f_1(x), \ldots, f_{n-1}(x)\}.$$

This function satisfies the conditions of the theorem and the inductive assumption. It follows that the manifold X is diffeomorphic to the Cartesian product $X = Z \times \mathbf{R}^{n-1}$ and the functions f_i are the coordinate functions for the second factor. Then grad f_n is tangent to the first factor Z and hence the function $f_n(x, t)$ does not depend on $t \in \mathbf{R}^{n-1}$. Therefore one can apply the first step of the induction to the manifold Z, that is,

$$Z = Z_1 \times \mathbf{R}^1.$$

Thus

$$X = Z_1 \times \mathbf{R}^n.$$

∎

7. An invariant formulation of the implicit function theorem Let $f : X \longrightarrow Y$ be a smooth mapping of smooth manifolds. The point $y_0 \in Y$ is called a *regular value of the mapping f* if for any point $x \in f^{-1}(y_0)$ the differential

$Df : T_x(X) {\longrightarrow} T_{y_0}(Y)$ is surjective. The implicit function theorem says that the inverse image

$$Z = f^{-1}(y_0)$$

is a smooth manifold and that

$$TX_{|Z} = TZ \oplus \mathbf{R}^m, \ m = \dim Y.$$

In general, if $W \subset Y$ is a submanifold then the mapping f is said to be *transversal along submanifold* W when for any point $x \in f^{-1}(W)$ one has

$$T_{f(x)}Y = T_{f(x)}W \oplus Df(T_x X).$$

In particular, the mapping f is transversal 'along' each regular point $y_0 \in Y$. The implicit function theorem says that the inverse image

$$Z = f^{-1}(W)$$

is a submanifold, the normal bundle $\nu(Z \subset X)$ is isomorphic to the bundle $f^*(\nu(W \subset Y))$ and the differential Df is the fiberwise isomorphism

$$\nu(Z \subset X) {\longrightarrow} \nu(W \subset Y).$$

8. *Morse functions on manifolds* Consider a smooth function f on a manifold X. A point p_0 is called a *critical point* if

$$df(x_0) = 0.$$

A critical point x_0 is said to be *nondegenerate* if the matrix of second derivatives is nondegenerate. This property does not depend on a choice of local coordinates. Let T^*X be the total space of the cotangent bundle of manifold X (that is, the vector bundle which is dual to the tangent bundle). Then for each function

$$f : X {\longrightarrow} \mathbf{R}^1$$

there is a mapping

$$df : X {\longrightarrow} T^*X \tag{1.74}$$

adjoint to Df which to each point $x \in X$ associates the linear form on $T_x X$ given by the differential of the function f at the point x. Then in the manifold T^*X there are two submanifolds: the zero section X_0 of the bundle T^*X and the image $df(X)$. The common points of these submanifolds correspond to critical points of the function f. Further, a critical point is nondegenerate if and only if the intersection of submanifolds X_0 and $df(X)$ at that point is transversal. If all the critical points of the function f are nondegenerate then

f is called a *Morse function*. Thus the function f is a Morse function if and only if the mapping (1.74) is transversal along the zero section $X_0 \subset T^*X$.

9. *Oriented manifolds.* A manifold X is said to be *oriented* if there is an atlas $\{U_\alpha\}$ such that all the transition functions $\varphi_{\alpha\beta}$ have positive Jacobians at each point. The choice of a such atlas is called an *orientation* of the manifold X. If the manifold X is orientable then the structure group $\mathbf{GL}\,(n,\ \mathbf{R})$ of the tangent bundle TX reduces to the subgroup $\mathbf{GL}^{\,+}(n,\ \mathbf{R})$ of matrices with positive determinant. Conversely, if the structure group $\mathbf{GL}\,(n,\ \mathbf{R})$ of the tangent bundle TX can be reduced to the subgroup $\mathbf{GL}^{\,+}(n,\ \mathbf{R})$, then the manifold X is orientable. Indeed, let $\{U_\alpha\}$ be an atlas,

$$\varphi_{\alpha\beta} = \left\| \frac{\partial x_\alpha^k}{\partial x_\beta^j} \right\|$$

be the transition functions of the tangent bundle and

$$\psi_{\alpha\beta} = h_\alpha \varphi_{\alpha\beta} h_\beta^{-1}$$

be the new transition functions such that

$$\det \psi_{\alpha\beta} > 0.$$

Of course, if the matrix valued functions

$$h_\alpha : U_\alpha \longrightarrow \mathbf{GL}\,(n,\ \mathbf{R})$$

had the form of the Jacobian matrix, that is, if

$$h\alpha(x) = \left\| \frac{\partial y_\alpha^k}{\partial x_\alpha^j} \right\|$$

then the functions

$$y_\alpha^k = y_\alpha^k(x_\alpha^1, \ldots, x_\alpha^n)$$

could serve as changes of coordinates giving an orientation on the manifold X. But in general this is not true and we need to find new functions h_α. Notice that if we have a suitable system of functions then one can change this system in any way that preserves the signs of $\det h_\alpha$. Hence, select new functions as follows:

$$\bar{h}_\alpha(x) = \left\| \begin{matrix} \pm 1 & 0 & \cdots & 0 \\ 0 & 1 & \cdots & 0 \\ \vdots & \vdots & \ddots & \vdots \\ 0 & 0 & \cdots & 1 \end{matrix} \right\|,$$

where the sign of the first entry should coincide with the sign of $\det h_\alpha(x)$ on each connected component of the chart U_α. Then the new transition functions

$$\bar{\psi}_{\alpha\beta}(x) = \bar{h}_\alpha(x)\varphi_{\alpha\beta}(x)\bar{h}_\beta^{-1}(x)$$

satisfy the same condition

$$\det \psi_{\alpha\beta}(x) > 0.$$

On the other hand, the functions $\bar{h}_\alpha(x)$ are the Jacobians of a change of the coordinates:

$$\begin{aligned}
y_\alpha^1 &= \pm x_\alpha^1, \\
y_\alpha^2 &= x_\alpha^2, \\
\cdots\ \cdots\ \cdots \\
y_\alpha^n &= x_\alpha^n.
\end{aligned}$$

and an atlas which gives an orientation of the manifold X has been found.

1.6 LINEAR GROUPS AND RELATED BUNDLES

This section consists of examples of vector bundles which arise naturally in connection with linear groups and their homogeneous spaces.

1.6.1 The Hopf bundle

The set of all one dimensional subspaces or lines of $\mathbf{R}^{n+1}$ is called the *real (n dimensional) projective space* and denoted by $\mathbf{RP}^n$. It has a natural topology given by the metric which measures the smaller angle between two lines. The projective space $\mathbf{RP}^n$ has the structure of smooth (and even real analytical) manifold. To construct the smooth manifold structure on $\mathbf{RP}^n$ one should first notice that each line l in $\mathbf{R}^{n+1}$ is uniquely determined by any nonzero vector x belonging to the line. Let $\{x_0, \ldots, x_n\}$ be the Cartesian coordinates of such a vector x, not all vanishing. Then the line l is defined by the coordinates $\{x_0, \ldots, x_n\}$ and any $\{\lambda x_0, \ldots, \lambda x_n\}$, $\lambda \neq 0$. Thus the point of $\mathbf{RP}^n$ is given by a class $[x_0 : x_1 : \ldots : x_n]$ of coordinates $\{x_0, \ldots, x_n\}$ (not all vanishing) determined up to multiplication by a nonzero real number λ. The class $[x_0 : x_1 : \ldots : x_n]$ gives the *projective coordinates* of the point of $\mathbf{RP}^n$.

We define an atlas $\{U_k\}_{k=0}^n$ on $\mathbf{RP}^n$ as follows. Put

$$U_k = \{[x_0 : x_1 : \ldots : x_n] : x_k \neq 0\},$$

and define coordinates on U_k by the following functions:

$$y_k^\alpha = \frac{x_\alpha}{x_k}, \; 0 \leq \alpha \leq n, \; \alpha \neq k.$$

where in the numbering of the coordinates by index α there is a gap when $\alpha = k$. The change of variables on the intersection $U_k \cap U_j$, $(k \neq j)$ of two charts has the following form:

$$y_k^\alpha = \begin{cases} \frac{y_j^\alpha}{y_j^k} & \text{when} \quad \alpha \neq j, \\ \frac{1}{y_j^k} & \text{when} \quad \alpha = j. \end{cases} \tag{1.75}$$

The formula (1.75) is well defined because $k \neq j$ and on $U_k \cap U_j$

$$y_j^k = \frac{x_k}{x_j} \neq 0.$$

All the functions in (1.75) are smooth functions making $\mathbf{RP}^n$ into a smooth manifold of dimension n.

Consider now the space E in which points have the form (l, x), where l is a one dimensional subspace of $\mathbf{R}^{n+1}$ and x is a point on l. The space E differs from the space $\mathbf{R}^{n+1}$ in that instead of zero vector of $\mathbf{R}^{n+1}$ in the space E there are many points of the type $(l, 0)$. The mapping

$$p : E \longrightarrow \mathbf{RP}^n, \tag{1.76}$$

which associates with each pair (l, x) its the first component, gives a locally trivial vector bundle. Indeed, the space E can be represented as a subset of the Cartesian product $\mathbf{RP}^n \times \mathbf{R}^{n+1}$ defined by the following system of equations in each local coordinate system:

$$\text{rank} \left\| \begin{matrix} x_0 & x_1 & \cdots & x_n \\ y_0 & y_1 & \cdots & y_n \end{matrix} \right\| = 1,$$

where $(x_0, x_1, \ldots, x_n)$ are projective coordinates of a point of $\mathbf{RP}^n$, and $(y_0, y_1, \ldots, y_n)$ are coordinates of the point of $\mathbf{R}^{n+1}$. For example, in the case when $U_0 = \{x_0 \neq 0\}$ we put $x_0 = 1$. Then

$$\text{rank} \left\| \begin{matrix} 1 & x_1 & \cdots & x_n \\ y_0 & y_1 & \cdots & y_n \end{matrix} \right\| = 1,$$

that is,

$$\begin{cases} \det \left\| \begin{matrix} 1 & x_k \\ y_0 & y_k \end{matrix} \right\| = 0, \\ \det \left\| \begin{matrix} x_k & x_j \\ y_k & y_j \end{matrix} \right\| = 0, \end{cases}$$

Hence the set E is defined in the $\mathbf{RP}^n \times \mathbf{R}^{n+1}$ by the following system of n equations:

$$f_k(x_1, \ldots, x_n, y_0, \ldots, y_n) = y_k - x_k y_0 = 0, \ k = 1, \ldots, n.$$

The Jacobian matrix of the functions f_k is

$$\left\| \begin{matrix} -y_0 & 0 & \ldots & 0 & -x_1 & 1 & 0 & \ldots & 0 \\ 0 & -y_0 & \ldots & 0 & -x_2 & 0 & 1 & \ldots & 0 \\ \ldots & \ldots & \ldots & \ldots & \ldots & \ldots & \ldots & \ldots & \ldots \\ 0 & 0 & \ldots & -y_0 & -x_n & 0 & 0 & \ldots & 1 \end{matrix} \right\| \tag{1.77}$$

Clearly the rank of this matrix (1.77) is maximal. By the implicit function theorem the space E is a submanifold of dimension $n+1$. As coordinates one can take $(y_0, x_1, \ldots, x_n)$. The projection (1.76) consists of forgetting the first coordinate y_0. So the inverse image $p^{-1}(U_0)$ is homeomorphic to the Cartesian product $U_0 \times \mathbf{R}^1$. Changing to another chart U_k, one takes another coordinate y_k as the coordinate in the fiber which depends linearly on y_0. Thus the mapping (1.76) gives a one dimensional vector bundle.

1.6.2   The complex Hopf bundle

This bundle is constructed similarly to the previous example as a one dimensional complex vector bundle over the base $\mathbf{CP}^n$. In the both cases, real and complex, the corresponding principal bundles with the structure groups $\mathbf{O}(1) = \mathbf{Z}_2$ and $\mathbf{U}(1) = \mathbf{S}^1$ can be identified with subbundles of the Hopf bundle. The point is that the structure group $\mathbf{O}(1)$ can be included in the fiber $\mathbf{R}^1$, $\mathbf{O}(1) = \{-1, 1\} \subset \mathbf{R}^1$, in a such way that the linear action of $\mathbf{O}(1)$ on $\mathbf{R}^1$ coincides on the subset $\{-1, 1\}$ with the left multiplication. Similarly, the group $\mathbf{U}(1) = \mathbf{S}^1$ can be included in $\mathbf{C}^1$:

$$\mathbf{S}^1 = \{z : |z| = 1\} \subset \mathbf{C}^1,$$

such that linear action of $\mathbf{U}(1)$ on $\mathbf{C}^1$ coincides on $\mathbf{S}^1$ with left multiplication. Hence the Hopf bundle has the principal subbundle consisting of vectors of unit length. Let

$$p : E_S \longrightarrow \mathbf{RP}^n$$

be principal bundle associated with the Hopf vector bundle. The points of the total space E_S are pairs (l, x), where l is a line in $\mathbf{R}_{n+1}$ and $x \in l, |x| = 1$. Since $x \neq 0$, the pair (l, x) is uniquely determined by the vector x. Hence the total space E_S is homeomorphic to the sphere $\mathbf{S}^n$ of unit radius a nd the principal bundle

$$p : \mathbf{S}^n \longrightarrow \mathbf{RP}^n$$

is the two-sheeted covering. In the case of the complex Hopf bundle, the associated principal bundle

$$p : E_S \longrightarrow \mathbf{CP}^n$$

is

$$E_S = \mathbf{S}^{2n+1}$$

and the fiber is a circle $\mathbf{S}^1$.

Both these two principal bundles are also called *Hopf bundles*. When $n = 1$, we have $\mathbf{S}^3 \longrightarrow \mathbf{P}^1$, that is, the classical Hopf bundle

$$\mathbf{S}^3 \longrightarrow \mathbf{S}^2. \tag{1.78}$$

In this last case it is useful to describe the transition functions for the intersection of charts. Let us consider the sphere $\mathbf{S}^3$ as defined by the equation

$$|z_0|^2 + |z_1|^2 = 1,$$

in the two dimensional complex vector space $\mathbf{C}^2$. The map (1.78) associates the point (z_0, z_1) to the point in the $\mathbf{CP}^1$ with the projective coordinates $[z_0 : z_1]$. So over the base $\mathbf{CP}^1$ we have the atlas consisting of two charts:

$$\begin{aligned} U_0 &= \{[z_0, z_1] : z_0 \neq 0\}, \\ U_1 &= \{[z_0, z_1] : z_1 \neq 0\}, \end{aligned}$$

The points of the chart U_0 are parametrized by the complex parameter

$$w_0 = \frac{z_1}{z_0} \in \mathbf{C}^1$$

whereas points of the chart U_1 are parametrized by the complex parameter

$$w_1 = \frac{z_0}{z_1} \in \mathbf{C}^1.$$

The homeomorphisms

$$\begin{aligned} \varphi_0 &: p^{-1}(U_0) \longrightarrow \mathbf{S}^1 \times \mathbf{C}^1 = \mathbf{S}^1 \times U_0, \\ \varphi_1 &: p^{-1}(U_1) \longrightarrow \mathbf{S}^1 \times \mathbf{C}^1 = \mathbf{S}^1 \times U_1, \end{aligned}$$

have the form

$$\varphi_0(z_0, z_1) = \left(\frac{z_0}{|z_0|}, \frac{z_1}{z_0}\right) = \left(\frac{z_0}{|z_0|}, [z_0 : z_1]\right),$$

$$\varphi_1(z_0, z_1) = \left(\frac{z_1}{|z_1|}, \frac{z_0}{z_1}\right) = \left(\frac{z_1}{|z_1|}, [z_0 : z_1]\right).$$

The mappings φ_0 and φ_1 clearly are invertible:

$$\varphi_0^{-1}(\lambda, w_0) = \left(\frac{\lambda}{\sqrt{1 + |w_0|^2}}, \frac{\lambda w_0}{\sqrt{1 + |w_0|^2}}\right),$$

$$\varphi_1^{-1}(\lambda, w_1) = \left(\frac{\lambda}{\sqrt{1 + |w_1|^2}}, \frac{\lambda w_1}{\sqrt{1 + |w_1|^2}}\right).$$

Hence the transition function

$$\varphi_{01} : \mathbf{S}_1 \times (U_0 \cap U_1) \longrightarrow \mathbf{S}_1 \times (U_0 \cap U_1)$$

is defined by the formula:

$$\varphi_{01}(\lambda, [z_{0:z_1}]) = \left(\frac{z_1 |z_0|}{z_0 |z_1|}, [z_0 : z_1],\right),$$

that is, multiplication by the number

$$\frac{z_1 |z_0|}{z_0 |z_1|} = \frac{w_1}{|w_1|}.$$

1.6.3   The tangent bundle of the Hopf bundle

Let ξ be the complex Hopf bundle over $\mathbf{CP}^n$ and let $T\mathbf{CP}^n$ be the tangent bundle. Then

$$T\mathbf{CP}^n \oplus \overline{1} = \xi^* \oplus \xi^* \oplus \ldots \oplus \xi^* = (n+1)\xi^*. \qquad (1.79)$$

When $n = 1$ this gives

$$\mathbf{TCP}^1 \oplus \overline{1} = \xi^* \oplus \xi^*.$$

Indeed, using the coordinate transition function

$$w_2 = \frac{1}{w_1},$$

the transition function for the tangent bundle TCP^1 has the following form:

$$\psi_{01}(w_1) = -\frac{1}{w_1^2} = -\frac{\bar{w}_1^2}{|w_1|^4}. \tag{1.80}$$

On the other hand, the transition function for the Hopf bundle is

$$\varphi_{01}(w_1) = \frac{w_1}{|w_1|}. \tag{1.81}$$

Using homotopies the formulas (1.80) and (1.81) can be simplified to

$$\psi_{01}(w_1) \;=\; \bar{w}_1^2,$$
$$\varphi_{01}(w_1) = w_1.$$

This shows that the matrix functions

$$\left\| \begin{matrix} \bar{w}_1^2 & 0 \\ 0 & 1 \end{matrix} \right\| \text{ and } \left\| \begin{matrix} \bar{w}_1 & 0 \\ 0 & \bar{w}_1 \end{matrix} \right\|$$

are homotopic in the class of invertible matrices when $w_1 \neq 0, w_1 \in \mathbf{C}^1$. In general, let us consider the manifold $\mathbf{CP}^n$ as the quotient space of the unit sphere $\mathbf{S}^{2n+1}$ in $\mathbf{C}^{n+1}$ by the action of the group $\mathbf{S}^1 \subset \mathbf{C}^1$ of complex numbers with unit norm. The total space of the tangent bundle TCP^n is the quotient space of the family of all tangent vectors of the sphere which are orthogonal to the orbits of the action of the group $\mathbf{S}^1$. In other words, one has

$$TCP^n = \frac{\{(u, v) : u, v \in \mathbf{C}^{n+1}, |u| = 1, \langle u, v \rangle = 0\}}{\{(\lambda u, \lambda v) \sim (u, v), \lambda \in \mathbf{S}^1\}},$$

where $\langle u, v \rangle$ is the Hermitian inner product in $\mathbf{C}^{n+1}$ Consider the quotient space

$$\mathbf{A} = \frac{\{(\bar{u}, s) : u \in \mathbf{C}^{n+1}, |\bar{u}| = 1, \ s \in \mathbf{C}^1\}}{\{(\lambda \bar{u}, \lambda s) \sim (\bar{u}, s), \lambda \in \mathbf{S}^1\}}.$$

Associate to each pair $(\bar{u}, s) \in \mathbf{A}$ the line l which passes through vector $\bar{u} \in \mathbf{C}^{n+1}$ and the vector $s\bar{u} \in l$. If $(\lambda \bar{u}, \lambda s)$ is an equivalent pair (which passes through vector $\lambda \bar{u}$) then it corresponds the same line l and the same vector $s\bar{u} = (\lambda s)(\overline{\lambda u})$. This means that the space $\mathbf{A}$ is homeomorphic to the total space of vector bundle ξ^*. Hence the total space of the vector bundle $(n+1)\xi^*$ is homeomorphic to the space

$$\mathbf{B} = \frac{\{(u, s) : u, v \in \mathbf{C}^{n+1}, |u| = 1, \}}{\{(\lambda u, \lambda v) \sim (u, v), \lambda \in \mathbf{S}^1\}}.$$

The space $TCP^n =$ is clearly a subspace of $\mathbf{B}$. A complementary subbundle can be defined as a quotient space

$$\mathbf{D} = \frac{\{(u,v) : u, v \in \mathbf{C}^{n+1}, |u| = 1, v = su, s \in \mathbf{C}^1\}}{\{(\lambda u, \lambda v) \sim (u,v), \lambda \in \mathbf{S}^1\}}.$$

The latter is homeomorphic to the space

$$\mathbf{D} = \frac{\{(u,s) : u \in \mathbf{C}^{n+1}, |u| = 1, s \in \mathbf{C}^1\}}{\{\lambda u \sim u, \lambda \in \mathbf{S}^1\}},$$

which is homeomorphic to the Cartesian product $TCP^n \times \mathbf{C}^1$. Thus one has the isomorphism (1.79).

1.6.4 Bundles of classical manifolds

Denote by $\mathbf{V}_{n,k}$ the space where the points are orthonormal sets of k vectors of n–dimensional euclidean space ($\mathbf{R}^n$ or $\mathbf{C}^n$). If we need we will write $\mathbf{V}_{n,k}^R$ or $\mathbf{V}_{n,k}^C$. Correspondingly, let us denote by $\mathbf{G}_{n,k}$ the space in which points are k–dimensional subspaces of n–dimensional euclidean space ($\mathbf{R}^n$ or $\mathbf{C}^n$). By expanding a k–frame to a basis in $\mathbf{C}^n$ we obtain

$$\mathbf{V}_{n,k}^C = \mathbf{U}(n)/\mathbf{U}(n-k),$$

where $\mathbf{U}(n-k) \subset \mathbf{U}(n)$ is a natural inclusion of unitary matrices:

$$A_{n-k} \longrightarrow \left\| \begin{matrix} 1_k & 0 \\ 0 & A_{n-k} \end{matrix} \right\|.$$

Similarly, the space $\mathbf{G}_{n,k}$ is homeomorphic to the homogeneous space

$$\mathbf{U}(n)/\left(\mathbf{U}(k) \oplus \mathbf{U}(n-k)\right),$$

where $\mathbf{U}(k) \oplus \mathbf{U}(n-k) \subset \mathbf{U}(n)$ is natural inclusion

$$(A_k, B_{n-k}) \longrightarrow \left\| \begin{matrix} A_k & 0 \\ 0 & B_{n-k} \end{matrix} \right\|.$$

Generally speaking, if G is a Lie group, and $H \subset G$ is a subgroup then the projection

$$p : G \longrightarrow G/H$$

is a locally trivial bundle (a principal H-bundle) since the rank of Jacobian matrix is maximal and consequently constant. Hence the following mappings give locally trivial bundles

$$\mathbf{V}_{n,k} \xrightarrow{\mathbf{V}_{n-k_1,k-k_1}} \mathbf{V}_{n,k_1},$$

$$\mathbf{V}_{n,k}^C \xrightarrow{\mathbf{U}(k)} \mathbf{G}_{n,k}^C.$$

(The fibers are shown over the arrows.) In particular,

$$\begin{aligned}
\mathbf{V}_{n,n}^R &= \mathbf{O}(n), \\
\mathbf{V}_{n,n}^C &= \mathbf{U}(n), \\
\mathbf{V}_{n,1}^R &= \mathbf{S}^{n-1}, \\
\mathbf{V}_{n,1}^C &= \mathbf{S}^{2n-1}.
\end{aligned}$$

Hence we have the following locally trivial bundles

$$\mathbf{U}(n) \xrightarrow{\mathbf{U}(n-1)} \mathbf{S}^{2n-1},$$

$$\mathbf{O}(n) \xrightarrow{\mathbf{O}(n-1)} \mathbf{S}^{n-1},$$

All the mappings above are defined by forgetting some of the vectors from the frame. The manifolds $\mathbf{V}_{n,k}$ are called the *Stieffel manifolds*, and the $\mathbf{G}_{n,k}$ called *Grassmann manifolds*.

2

HOMOTOPY INVARIANTS OF
VECTOR BUNDLES

In the section 1.2 of the chapter 1 (Theorem 3) it was shown that if the base of a locally trivial bundle with a Lie group as structure group has the form $B \times I$ then the restrictions of the bundle to $B \times \{0\}$ and $B \times \{1\}$ are isomorphic. This property of locally trivial bundles allows us to describe bundles in the terms of homotopy properties of topological spaces.

2.1   THE CLASSIFICATION THEOREMS

Assume that the base B of a locally trivial bundle

$$p : E \longrightarrow B$$

is a cellular space, that is, the space B is a direct limit of its compact subspaces $[B]^n$,

$$B = \lim_{n \to \infty} [B]^n,$$

and each space $[B]^n$ is constructed from $[B]^{(n-1)}$ by attaching to it a finite family of discs D_i^n by using continuous mappings

$$\mathbf{S}_i^{n-1} \longrightarrow [B]^{(n-1)}.$$

The subspace $[B]^n$ is called the *n–dimensional skeleton* of the space B.

Let us consider a principal G-bundle,

$$p_G : E_G \longrightarrow B_G, \qquad\qquad (2.1)$$

such that all homotopy groups of the total space E_G are trivial:

$$\pi_i(E_G) =), \; 0 \leq i < \infty. \tag{2.2}$$

Theorem 13 *Let B be a cellular space. Then any principal G-bundle*

$$p : E \longrightarrow B \tag{2.3}$$

is isomorphic to the inverse image of the bundle (2.1) with respect to a continuous mapping

$$f : B \longrightarrow B_G. \tag{2.4}$$

Two inverse images of the bundle (2.1) with respect to the mappings

$$f, g : B \longrightarrow B_G$$

are isomorphic if and only if the mappings f and g are homotopic.

Corollary 2 *The family of all isomorphism classes of principal G-bundles over the base B is in one to one correspondence with the family of homotopy classes of continuous mappings from B to B_G.*

Corollary 3 *If two cellular spaces B and B' are homotopy equivalent then the families of all isomorphism classes of principal G-bundles over the bases B and B' are in one to one correspondence. This correspondence is defined by inverse image with respect to a homotopy equivalence*

$$B \longrightarrow B'.$$

Corollary 4 *If the space B_G is a cellular space then it is defined by the condition (2.2) up to homotopy equivalence.*

Proof of Theorem 13.

By Theorem 2 of section 1.2 of chapter 1 the bundle (2.3) is isomorphic to an inverse image with respect to a continuous mapping (2.4) if and only if there is an equivariant mapping of total spaces of these bundles. Let B be a cellular space, B_0 be a cellular subspace, (that is, B_0 is union of some cells of the space B and is closed subspace). Consider a principal bundle (2.3) and let

$$E_0 = p^{-1}(B_0).$$

Let

$$F_0 : E_0 \longrightarrow E_G \qquad (2.5)$$

be an equivariant continuous mapping. Let us show that F_0 can be extended to a continuous equivariant mapping

$$F : E \longrightarrow E_G. \qquad (2.6)$$

Lemma 3 *Any principal G–bundle p over a disc D^n is isomorphic to a trivial bundle.*

Proof.

In fact, put

$$h : D^n \times I \longrightarrow D^n,$$
$$h(x,t) = tx, \ 0 \le x \le 1, \ x \in D^n.$$

Then

$$h(x,0) \equiv x$$
$$h(x,1) \equiv 0.$$

Then by Theorem 3 of the section 1.2 of the chapter 1, inverse images of the bundle p with respect to the mappings

$$h_0(x) = h(x,0),$$
$$h_1(x) = h(x,1)$$

are isomorphic. On the other hand, the inverse image of the bundle p with respect to the identity mapping h_1 is isomorphic to the bundle p and the inverse image with respect to the mapping h_0 is trivial. ∎

Returning to the proof of Theorem 13, let the $(n-1)$–dimensional skeleton $[B]^{(n-1)}$ lie in the subspace B_0 with a cell D_i^n outside B_0. Let

$$\varphi_i : D_i^n \longrightarrow B$$

be the natural inclusion. Then

$$\varphi_i(\mathbf{S}_i^{(n-1)}) \subset [B]^{(n-1)} \subset B_0. \tag{2.7}$$

Consider the restriction of the bundle p to the cell D_i^n. By Lemma 3 this restriction is a trivial bundle, that is, the total space is isomorphic to $D_i^n \times G$. By (2.7) the mapping (2.5) induces the equivariant mapping

$$(F_0)_{|(\mathbf{S}_i^{(n-1)} \times G)} : (\mathbf{S}_i^{(n-1)} \times G) \longrightarrow E_G$$

$$F_0(x,g) = F_0(x,1)g, \qquad x \in \mathbf{S}_i^{(n-1)}, \ g \in G.$$

Consider the mapping

$$F_0(x,1) : \mathbf{S}_i^{(n-1)} \longrightarrow E_G. \tag{2.8}$$

Since

$$\pi_{(n-1)}(E_G) = 0,$$

the mapping (2.8) can be extended to a continuous mapping

$$F(x,1) : D_i^n \longrightarrow E_G.$$

Put

$$F(x,g) = F(x,1)g, \ x \in D_i^n, \ g \in G. \tag{2.9}$$

It is clear that the mapping (2.9) is equivariant and extends the mapping F_0 to the set $p^{-1}(B_0 \cup D_i^n) \supset E_0$. Using induction on the number of cells we can establish the existence of an equivariant mapping (2.6) which extends the mapping (2.5). The first statement of the theorem follows by putting $B_0 = \emptyset$. The second statement of the theorem follows if we substitute B for $B \times I$ and put

$$B_0 = (B \times \{0\}) \cup (B \times \{1\}).$$

∎Corollaries 2 and 3 follow directly from Theorem 13.

Proof of Corollary 4.

Let

$$p : E_G \longrightarrow B_G,$$
$$p' : E'_G \longrightarrow B'_G$$

be two principal $G-$ bundles such that

$$\pi_i(E_G) = \pi_i(E'_G) = 0, \ 0 \le i \le \infty,$$

and let B_G and B_G' be cellular spaces. Then by Theorem 13 there are two continuous mappings

$$f : B_G \longrightarrow B_G',$$
$$g : B_G' \longrightarrow B_G$$

such that

$$f^*(p') = p,$$
$$g^*(p) = p'.$$

Hence

$$(gf)^*(p) = p = (\mathbf{Id})^*(p),$$
$$(gf)^*(p') = p' = (\mathbf{Id})^*(p').$$

By the second part of Theorem 13,

$$gf \sim \mathbf{Id}.$$
$$fg \sim \mathbf{Id}.$$

This means that the spaces B_G and B_G' are homotopy equivalent. ∎

The bundle (2.1) is called the *universal principal G-bundle* and the base B_G is called the *classifying space* for bundles with structure group G

2.2 EXACT HOMOTOPY SEQUENCE

It was shown in Theorem 13 of section 2.1 that the homotopy groups of the total space total space are needed for the construction of a universal principal G–bundle.

Definition 4 *A mapping*

$$p : X \longrightarrow Y$$

is said to satisfy the lifting homotopy axiom if for any cellular space P, any homotopy

$$f : P \times I \longrightarrow Y$$

and any mapping

$$g_0 : P \times \{0\} \longrightarrow Y$$

satisfying the condition
$$pg_0(x, 0) = f(x, 0),$$

there is an extension
$$g : P \times I \longrightarrow Y$$

satisfying a similar condition

$$pg(x, t) = f(x, t), \ x \in P, \ t \in I. \tag{2.10}$$

The mapping g is a lifting mapping (for the mapping f) with respect to the mapping p if the two mappings f and g satisfy the condition (2.10). So the lifting homotopy axiom states that if the initial mapping $f(x, 0)$ has a lifting mapping then any homotopy $f(x, t)$ has a lifting homotopy.

Theorem 14 *Let*
$$p : E \longrightarrow B \tag{2.11}$$
be a locally trivial bundle. Then p satisfies the lifting homotopy axiom.

Proof.

Assume firstly that (2.11) is a trivial bundle, that is,

$$E = B \times F.$$

Then the lifting mapping g_0 can be described as

$$g_0(x, 0) = (f(x, 0), h(x)).$$

Then let
$$g(x, t) = (f(x, t), h(x)). \tag{2.12}$$

Clearly the mapping (2.12) is a lifting mapping for the mapping f. Further, let $P_0 \subset P$ be a cellular subspace with a lifting homotopy

$$g_0 : P_0 \times I \longrightarrow E = B \times F.$$

In other words, the lifting mapping is defined on the subspace

$$(P_0 \times I) \cup (P \times \{0\}) \subset P \times I.$$

We need to extend the mapping g_0 to a lifting mapping

$$g : P \times I \longrightarrow E.$$

Notice that it is sufficient to be able to construct an extension of the mapping g_0 on a single cell

$$D_i^k \times I \subset P \times I.$$

The restriction of the mapping g_0 defines the lifting mapping on the subset

$$(D_i^k \times \{0\}) \cup (\mathbf{S}_i^{(k-1)} \times I) \subset D_i^k \times I. \tag{2.13}$$

This subset has the form of a 'glass' where the 'bottom' corresponds to $D_i^k \times \{0\}$ and the 'sides' to $\mathbf{S}_i^{(k-1)} \times I$. So the Cartesian product $D_i^k \times I$ is homeomorphic to itself in such a way that the subset (2.13) lies on the 'bottom'. Hence the problem of the construction of an extension of the mapping g_0 from (2.13) to the whole Cartesian product is equivalent to the problem of an extension of the lifting mapping from the 'bottom' $D_i^k \times \{0\}$ to the $D_i^k \times I$ and the latter problem is already solved. To complete the proof, consider a sufficiently fine cell decomposition of the space

$$P = \cup D_i^k$$

and the interval

$$I = \cup I_l$$

such that image of each Cartesian product of a cell with an interval lies in a chart of the bundle p. Then the restriction of the mapping f onto $D_i^k \times I_l$ maps into a chart U and the lifting mapping g maps it into the total space of the bundle $p_{|p^{-1}(U)}$. The latter is a trivial bundle. Hence the problem is reduced to the case we have proved. ∎

Theorem 15 *Let*

$$p : E \longrightarrow B$$

be a locally trivial bundle, $x_0 \in B$, $y_0 \in p^{-1}(x_0) = F$ and let

$$j : F \longrightarrow E$$

be a natural inclusion of the fiber in the total space. Then there is a homomorphism of homotopy groups

$$\partial : \pi_k(B, x_0) \longrightarrow \pi_{k-1}(F, y_0)$$

such that the sequence of homomorphisms

$$\ldots \longrightarrow \pi_k(F, y_0) \xrightarrow{j_*} \pi_k(E, y_0) \xrightarrow{p_*} \pi_k(B, x_0) \xrightarrow{\partial}$$
$$\xrightarrow{\partial} \pi_{k-1}(F, y_0) \xrightarrow{j_*} \pi_{k-1}(E, y_0) \xrightarrow{p_*} \pi_{k-1}(B, x_0) \xrightarrow{\partial} \ldots$$
$$\ldots \longrightarrow \pi_1(F, y_0) \xrightarrow{j_*} \pi_1(E, y_0) \xrightarrow{p_*} \pi_1(B, x_0) \tag{2.14}$$

is an exact sequence of groups.

Proof.

The elements of homotopy group $\pi_k(X, x_0)$ can be represented as homotopy classes of continuous mappings

$$\varphi : I^k \longrightarrow X$$

such that

$$\varphi(\partial I^k) = x_0.$$

To construct the mapping ∂ let

$$\varphi : I^k \longrightarrow B$$

be a representative of an element of the group $\pi_k(B, x_0)$ and note that

$$I^k = I^{k-1} \times I.$$

Hence

$$\varphi(I^{k-1} \times \{0\}) = \varphi(I^{k-1} \times \{1\}) = x_0.$$

The mapping φ can be considered as a homotopy of the cube I^{k-1}. Put

$$\psi(u, 0) \equiv y_0, \ u \in I^{k-1}.$$

Then the mapping ψ is a lifting of the mapping $\varphi(u, 0)$. By the lifting homotopy axiom there is a lifting homotopy

$$\psi : I^{k-1} \times T \longrightarrow E,$$

such that

$$p\psi(u, t) \equiv \varphi(u, t).$$

In particular,

$$p\psi(u, 1) \equiv x_0.$$

This means that $\psi(u, 1)$ maps the cube I^{k-1} into the fiber F. Moreover, since

$$\varphi(\partial I^{k-1} \times I) = x_0$$

the lifting ψ can be chosen such that

$$\psi(\partial I^{k-1} \times I) \equiv y_0.$$

Hence the mapping $\psi(u, 1)$ maps ∂I^{k-1} to the point y_0. This means that the mapping $\psi(u, 1)$ defines an element of the group $\pi_{k-1}(F, y_0)$. This homomorphism ∂ is well defined. Indeed, if the mapping φ is homotopic to the mapping φ' then the corresponding lifting mappings ψ and ψ' are homotopic and a homotopy Ψ between ψ and ψ' can be constructed as a lifting mapping for the homotopy Φ between φ and φ'. Hence the homomorphism ∂ is well defined. The proof of additivity of ∂ is left to the reader. Next we need to prove the exactness of the sequence (2.14).

1. Exactness at the term $\pi_k(E, y_0)$.

If

$$\varphi : \mathbf{S}^k \longrightarrow F$$

represents an element of the group $\pi_k(F, y_0)$, then the mapping

$$pj\varphi : \mathbf{S}^k \longrightarrow B$$

represents the image of this element with respect to homomorphism $p_* j_*$. Since

$$p(F) = x_0,$$

we have

$$pj\varphi(\mathbf{S}^k) = x_0.$$

Hence

$$\mathbf{Im}\ j_* \subset \mathbf{Ker}\ p_*.$$

Conversely, let

$$\varphi : \mathbf{S}^k \longrightarrow E$$

be a mapping such that $p\varphi$ is homotopic to a constant, that is,

$$[\varphi] \in \mathbf{Ker}\ p_*.$$

Let Φ be a homotopy between $p\varphi$ and a constant mapping. By the lifting homotopy axiom there is a lifting homotopy Ψ between φ and a mapping φ' which is lifting of the constant mapping. Thus

$$p\varphi'(\mathbf{S}^k) = x_0,$$

that is,

$$\varphi'(\mathbf{S}^k) \subset F.$$

Hence
$$[\varphi'] = [\varphi] \in \mathbf{Im}\ j_*,$$

that is,
$$\mathbf{Ker}\ p_* \subset \mathbf{Im}\ j_*.$$

2. Exactness in the term $\pi_k(B, y_0)$.

Let
$$\varphi : I^k \longrightarrow E$$

represent an element of the group $\pi_k(B, x_0)$. The composition

$$p\varphi : I^k = I^{k-1} \times I \longrightarrow B$$

is lifted by φ. According to definition, the restriction

$$\varphi(u, 1) : I^{k-1} \longrightarrow f \subset E$$

represents the element $\partial[p\varphi]$. Since

$$\varphi(u, 1) \equiv y_0,$$

we have
$$\partial p_*[\varphi] = 0.$$

Now let
$$\varphi : I_k = I_{k-1} \times I \longrightarrow B$$

represent an element of the group $\pi_k(B, x_0)$ and

$$\psi : I^{k-1} \times I \longrightarrow E$$

be a lifting of the mapping φ. Then the restriction $\psi(u, 1)$ represents the element $\partial[\varphi]$. If the mapping

$$\psi(u, 1) : I^{k-1} \longrightarrow F$$

is homotopic to the constant mapping then there is a homotopy

$$\psi' : I^{k-1} \times I \longrightarrow F$$

between $\psi(u, 1)$ and constant mapping. Let us now construct a new mapping

$$\psi'' : I^{k-1} \times I \longrightarrow E,$$

$$\psi''(u,1) = \begin{cases} \psi(u,2t), & 0 \le t \le \frac{1}{2} \\ \psi'(u,2t-1), & \frac{1}{2} \le t \le 1. \end{cases}$$

Then

$$p\psi''(u,1) = \begin{cases} \varphi(u,2t), & 0 \le t \le \frac{1}{2} \\ x_0, & \frac{1}{2} \le t \le 1. \end{cases}$$

Hence the mapping $p\psi''$ is homotopic to the mapping φ, that is,

$$p_*[\psi''] = [\varphi]$$

or

$$\mathbf{Ker}\ \partial \subset \mathbf{Im}\ p_*.$$

3. Exactness in the term $\pi_k(F,y_0)$.

Let the mapping

$$\varphi : I^{k+1} = I^k \times I \longrightarrow B$$

represent an element of the group $\pi_{k+1}(B,x_0)$ and let

$$\psi : I^k \times I \longrightarrow E$$

lift the mapping φ. Then the restriction $\psi(u,1)$ represents the element

$$\partial([\varphi]) \in \pi_k(F,y_0).$$

Hence the element $j_*\partial([\varphi])$ is represented by the mapping $\psi(u,1)$. The latter is homotopic to constant mapping $\psi(u,0)$, that is,

$$j_*\partial([\varphi]) = 0.$$

Conversely, if

$$\varphi : I^k \times I \longrightarrow E$$

is a homotopy between $\varphi(u,1) \subset F$ and constant mapping $\varphi(u,0)$, then according to the definition the mapping $\varphi(u,1)$ represents the element $\partial([p\varphi])$. Hence

$$\mathbf{Ker}\ j_* \subset \mathbf{Im}\ \partial.$$

∎

Remark

Let $\pi_0(X, x_0)$ denote the set of connected components of the space X where the fixed element $[x_0]$ is the component containing the point x_0. Then the exact sequence (2.14) can be extended as follows

$$\ldots \longrightarrow \pi_1(F, y_0) \xrightarrow{j_*} \pi_1(E, y_0) \xrightarrow{p_*} \pi_1(B, x_0) \xrightarrow{\partial}$$
$$\longrightarrow \pi_0(F, y_0) \xrightarrow{j_*} \pi_0(E, y_0) \xrightarrow{p_*} \pi_0(B, x_0).$$

This sequence is exact in the sense that the image coincides with the inverse image of the fixed element.

Examples

1. Consider the covering

$$\mathbf{R}^1 \longrightarrow \mathbf{S}^1$$

in which the fiber $\mathbf{Z}$ is the discrete space of integers. The exact homotopy sequence

$$\ldots \longrightarrow \pi_k(\mathbf{Z}) \xrightarrow{j_*} \pi_k(\mathbf{R}^1) \xrightarrow{p_*} \pi_k(\mathbf{S}^1) \xrightarrow{\partial} \ldots$$
$$\ldots \xrightarrow{\partial} \pi_1(\mathbf{Z}) \xrightarrow{j_*} \pi_1(\mathbf{R}^1) \xrightarrow{p_*} \pi_1(\mathbf{S}^1) \xrightarrow{\partial}$$
$$\longrightarrow \pi_0(\mathbf{Z}) \xrightarrow{j_*} \pi_0(\mathbf{R}^1) \xrightarrow{p_*} \pi_0(\mathbf{S}^1)$$

has the following form

$$\ldots \longrightarrow 0 \xrightarrow{j_*} 0 \xrightarrow{p_*} \pi_k(\mathbf{S}^1) \xrightarrow{\partial} 0 \longrightarrow \ldots$$
$$\ldots \xrightarrow{\partial} 0 \xrightarrow{j_*} 0 \xrightarrow{p_*} \pi_1(\mathbf{S}^1) \xrightarrow{\partial}$$
$$\longrightarrow \mathbf{Z} \xrightarrow{j_*} 0 \xrightarrow{p_*} 0,$$

since

$$\pi_k(\mathbf{R}^1) = 0, \qquad k \geq 0,$$
$$\pi_k(\mathbf{Z}) = 0, \qquad k \geq 1.$$

Hence

$$\pi_1(\mathbf{S}^1) \;\; = \;\; \mathbf{Z},$$
$$\pi_k(\mathbf{S}^1) \;\; = \;\; 0 \quad \text{when } k \geq 2.$$

2. Consider the Hopf bundle

$$p : \mathbf{S}^3 \longrightarrow \mathbf{S}^2.$$

The corresponding exact homotopy sequence

$$\ldots \longrightarrow \pi_k(\mathbf{S}^1) \xrightarrow{j_*} \pi_k(\mathbf{S}^3) \xrightarrow{p_*} \pi_k(\mathbf{S}^2) \xrightarrow{\partial} \ldots$$
$$\ldots \xrightarrow{\partial} \pi_3(\mathbf{S}^1) \xrightarrow{j_*} \pi_3(\mathbf{S}^3) \xrightarrow{p_*} \pi_3(\mathbf{S}^2) \xrightarrow{\partial}$$
$$\ldots \xrightarrow{\partial} \pi_2(\mathbf{S}^1) \xrightarrow{j_*} \pi_2(\mathbf{S}^3) \xrightarrow{p_*} \pi_2(\mathbf{S}^2) \xrightarrow{\partial}$$
$$\ldots \xrightarrow{\partial} \pi_1(\mathbf{S}^1) \xrightarrow{j_*} \pi_1(\mathbf{S}^3) \xrightarrow{p_*} \pi_1(\mathbf{S}^2) \xrightarrow{\partial}$$

has the following form:

$$\ldots \longrightarrow 0 \xrightarrow{j_*} \pi_k(\mathbf{S}^3) \xrightarrow{p_*} \pi_k(\mathbf{S}^2) \xrightarrow{\partial} 0 \; dots$$
$$\ldots \xrightarrow{\partial} 0 \xrightarrow{j_*} \mathbf{Z} \xrightarrow{p_*} \pi_3(\mathbf{S}^2) \xrightarrow{\partial}$$
$$\ldots \xrightarrow{\partial} 0 \xrightarrow{j_*} 0 \xrightarrow{p_*} \pi_2(\mathbf{S}^2) \xrightarrow{\partial}$$
$$\ldots \xrightarrow{\partial} \mathbf{Z} \xrightarrow{j_*} 0 \xrightarrow{p_*} 0,$$

since

$$\pi_3(\mathbf{S}^1) = \pi_2(\mathbf{S}^1) = \pi_2(\mathbf{S}^3) =$$
$$= \pi_1(\mathbf{S}^3) = \pi_1(\mathbf{S}^2) = 0.$$

Hence

$$\pi_2(\mathbf{S}^2) = \pi_1(\mathbf{S}^1) = \mathbf{Z},$$
$$\pi_3(\mathbf{S}^2) = \mathbf{Z}.$$

The space $\mathbf{S}^2$ is the simplest example of a space where there are nontrivial homotopy groups in degrees greater than the dimension of the space. Notice also that the Hopf bundle projection $p : \mathbf{S}^3 \longrightarrow \mathbf{S}^2$ gives a generator of the group $\pi_3(\mathbf{S}^2) = \mathbf{Z}$. Indeed, from the exact homotopy sequence we see that a generator of the group $\pi_3(\mathbf{S}^3)$ is mapped to a generator of the group $\pi_3(\mathbf{S}^2)$ by the homomorphism p_*. A generator of the group $\pi_3(\mathbf{S}^3)$ is represented by the identity $\varphi : \mathbf{S}^3 \longrightarrow \mathbf{S}^3$ and hence

$$p_*[\varphi] = [p\varphi] = [p].$$

3. In the general case there are two bundles: the real Hopf bundle

$$\mathbf{S}^n \longrightarrow \mathbf{RP}^n$$

with fiber $\mathbf{Z}_2$, and complex Hopf bundle

$$\mathbf{S}^{2n-1} \longrightarrow \mathbf{CP}^n$$

with fiber S^1. In the first case, the exact homotopy sequence shows that

$$\pi_1(S^n) = 0 \longrightarrow \pi_1(RP^n) \longrightarrow \pi_0(Z_2) = Z_2 \longrightarrow 0, \qquad \pi_1(RP^n) = Z_2,$$
$$\pi_k(S^n) = 0 \longrightarrow \pi_k(RP^n) \longrightarrow \pi_{k-1}(Z_2) = 0, \qquad 2 \le k \le n.$$

Hence

$$\pi_k(RP^n) = 0, \quad 2 \le k \le n-1.$$

When $n = 1$, we have $RP^1 = S^1$.
In the complex case, one has

$$0 = \pi_2(S^{2n-1}) \longrightarrow \pi_2(CP^n) \longrightarrow \pi_1(S^1) = Z \qquad \longrightarrow$$
$$\longrightarrow \pi_1(S^{2n-1}) = 0 \longrightarrow \pi_1(CP^n) \longrightarrow \pi_0(S^1) = 0$$
$$\pi_k(S^{2n-1}) = 0 \longrightarrow \pi_k(CP^n) \longrightarrow \pi_{k-1}(S^1) = 0$$

where $2n - 1 > k > 2$. Hence

$$\begin{aligned}
\pi_1(CP^n) &= 0, \\
\pi_2(CP^n) &= Z, \\
\pi_k(CP^n) &= 0 \text{ for } 3 \le k < 2n - 1.
\end{aligned}$$

4. Consider natural inclusion of matrix groups

$$j : O(n) \longrightarrow 0(n+1).$$

In previous chapter it was shown that j is a fiber of the bundle

$$p : O(n+1) \longrightarrow S^n.$$

So we have the exact homotopy sequence

$$\dots \longrightarrow \pi_{k+1}(S^n) \longrightarrow \pi_k(O(n)) \overset{j_*}{\longrightarrow} \pi_k(O(n+1)) \longrightarrow \pi_k(S^n) \longrightarrow \dots \qquad (2.15)$$

If $k + 1 < n$ that is $k \le n - 2$ then

$$\pi_k(O(n)) \equiv \pi_k(O(n+1)).$$

Hence

$$\pi_k(O(k+2)) \equiv \pi_k(O(k+3)) \equiv \pi_k(O(N))$$

for $N \ge k + 2$. This property is called the *stability of homotopy groups of the series of orthogonal groups*. In particular, when $k = 1$ one has

$$\begin{aligned}
\pi_1(O(1)) &= 0, \\
\pi_1(O(2)) &= Z, \\
\pi_1(O(3)) &= \pi_1(O(4)) = \dots = \pi_1(O(N)).
\end{aligned}$$

Also $SO(3) = RP^3$, and so

$$\pi_1(O(3) = \pi_1(RP^3) = Z_2.$$

Hence

$$\pi_1(O(n)) = Z_2 \text{ for } n \geq 3.$$

Moreover from (2.15) one has the exact sequence

$$Z = \pi_1(O(2)) \longrightarrow \pi_1(O(3)) = Z_2 \longrightarrow 0.$$

This means that the inclusion $O(2) \subset O(3)$ induces epimorphism of fundamental groups.

5. Consider the inclusion in the complex case

$$j : U(n) \longrightarrow U(n+1)$$

giving a fiber of the bundle

$$p : U(n+1) \longrightarrow S^{2n+1}.$$

The exact homotopy sequence

$$\dots \longrightarrow \pi_{k+1}(S^{2n+1}) \longrightarrow \pi_k(U(n)) \xrightarrow{j_*} \pi_k(U(n+1)) \longrightarrow \pi_k(S^{2k+1}) \longrightarrow \dots \quad (2.16)$$

shows that

$$\pi_k(U(n)) = \pi_k(U(n+1)) \text{ for } k < 2n.$$

and, in particular,

$$\pi_1(U(1)) = \dots = \pi_1(U(n)).$$

From $U(1) = S^1$ it follows that

$$\pi_1(U(n)) = Z, \quad n \geq 1.$$

For $k = 2$,

$$\pi_2(U(2)) = \dots = \pi_2(U(n)).$$

From (2.16) it follows that

$$\dots \longrightarrow p_3(S^3) \longrightarrow \pi_2(U(1)) \longrightarrow \pi_2(U(2)) \longrightarrow \pi_2(S^3) \longrightarrow \dots,$$

and hence

$$0 \longrightarrow \pi_2(U(2)) \longrightarrow 0,$$

or

$$\pi_2(U(2)) = 0.$$

Hence
$$\pi_2(\mathbf{U}(n)) = 0 \text{ for any } n.$$
Let us now calculate $\pi_3(\mathbf{U}(n))$. Firstly, for $n \geq 2$,
$$\pi_3(\mathbf{U}(2)) = \ldots = \pi_3(\mathbf{U}(n)).$$
and so it is sufficient to calculate the group $\pi_3(\mathbf{U}(2))$. Consider the locally trivial bundle
$$p : \mathbf{U}(2) \longrightarrow \mathbf{S}^1, \tag{2.17}$$
where $p(A) = \det A$. The fiber of (2.17) is the group $\mathbf{SU}(2)$. Thus we have the exact homotopy sequence
$$\ldots \longrightarrow \pi_4(\mathbf{S}^1) \longrightarrow \pi_3(\mathbf{SU}(2)) \longrightarrow \pi_3(\mathbf{U}(2)) \longrightarrow \pi_3(\mathbf{S}^1) \longrightarrow \ldots.$$
The end terms are trivial and hence
$$\pi_3(\mathbf{SU}(2)) = \pi_3(\mathbf{U}(2)).$$
Further, the group $\mathbf{SU}(2)$ is homeomorphic to the 3–dimensional sphere $\mathbf{S}^3 \subset \mathbf{C}^2$, and therefore
$$\pi_3(\mathbf{SU}(2)) = Z.$$
Thus,
$$\begin{aligned}
\pi_3(U(1)) &= \mathbf{0}, \\
\pi_3(U(n)) &= \mathbf{Z}, \quad n \geq 2.
\end{aligned}$$

6. In the same spirit as the two previous examples consider locally trivial bundle
$$p : \mathbf{Sp}(n) \longrightarrow \mathbf{S}^{4n-1}$$
with a fiber $\mathbf{Sp}(n-1)$ where $\mathbf{Sp}(n)$ is the group of orthogonal quaternionic transformations of the quaternionic space $\mathbf{K}^n = R^{4n}$. Each such transformation can be described as a matrix with entries which are quaternions. Hence each column is a vector in $\mathbf{K}^n = R^{4n}$. The mapping p assigns the first column to each matrix. The exact homotopy sequence
$$\ldots \longrightarrow \pi_k(\mathbf{Sp}(n-1)) \longrightarrow \pi_k(\mathbf{Sp}(n)) \longrightarrow \pi_k(\mathbf{S}^{4n-1}) \longrightarrow \ldots$$
$$\ldots \longrightarrow \pi_1(\mathbf{Sp}(n-1)) \longrightarrow \pi_1(\mathbf{Sp}(n)) \longrightarrow \pi_1(\mathbf{S}^{4n-1}) \longrightarrow 0$$
gives the following table of isomorphisms
$$\begin{aligned}
\pi_1(\mathbf{Sp}(n)) = \pi_1(\mathbf{Sp}(1)) = \pi_1(\mathbf{S}^3) &= \mathbf{0}, \\
\pi_2(\mathbf{Sp}(n)) = \pi_2(\mathbf{Sp}(1)) = \pi_2(\mathbf{S}^3) &= \mathbf{0}, \\
pi_3(\mathbf{Sp}(n)) = \pi_3(\mathbf{Sp}(1)) = \pi_3(\mathbf{S}^3) &= \mathbf{Z}, \\
\pi_k(\mathbf{Sp}(n)) = \pi_k(\mathbf{Sp}(n-1)) & \qquad \text{for } k < 4n - 2.
\end{aligned}$$

7. Let

$$p : E_G \longrightarrow B_G$$

be a universal principal G–bundle. By the exact homotopy sequence

$$\ldots \longrightarrow \pi_k(E_G) \longrightarrow \pi_{k-1}(B_G) \longrightarrow \pi_{k-1}(G) \longrightarrow \pi_{k-1}(E_G) \longrightarrow \ldots$$

we have

$$\pi_k(B_G) \equiv \pi_{k-1}(G)$$

since $\pi_k(E_G) = 0$ for all $k \geq 0$.

8. Let $G = \mathbf{S}^1$. Then the space $B_{\mathbf{S}^1}$ can be constructed as a direct limit

$$\mathbf{CP}^\infty = \lim \mathbf{CP}^n,$$

and total space $E_{\mathbf{S}^1}$ the direct limit

$$\mathbf{S}^\infty = \lim \mathbf{S}^{2n+1}.$$

Therefore,

$$\begin{aligned} \pi_2(\mathbf{CP}^\infty) &= \mathbf{Z} \\ \pi_k(\mathbf{CP}^\infty) &= 0 \text{ for } k \neq 2. \end{aligned} \tag{2.18}$$

The table (2.18) means that the space $\mathbf{CP}^\infty$ is an Eilenberg–McLane complex $K(\mathbf{Z}, 2)$. Thus the same space $\mathbf{CP}^\infty$ classifies both the one–dimensional complex vector bundles and two–dimensional integer cohomology group. Hence the family **Line** (X) of one dimensional complex vector bundles over the base X is isomorphic to the cohomology group $H^2(X, \mathbf{Z})$. Clearly, there is a one to one correspondence

$$l : \mathbf{Line}\ (X) \longrightarrow [X, \mathbf{CP}^\infty]$$

defined by the formula

$$l([f]) = f^*(\xi),$$

where $f : X \longrightarrow \mathbf{CP}^\infty$ is a continuous mapping, $[f] \in [X, \mathbf{CP}^\infty]$, and ξ is universal one–dimensional vector bundle over $\mathbf{CP}^\infty$. Then the isomorphism

$$\alpha : \mathbf{Line}\ (X) \longrightarrow H^2(X, \mathbf{Z}),$$

mentioned above is defined by the formula

$$\alpha(\eta) = \left(l^{-1}(\eta)\right)^*(a) \in H^2(X, \mathbf{Z}), \ \eta \in \mathbf{Line}\ (X),$$

where $a \in H^2(\mathbf{CP}^\infty, \mathbf{Z})$ is a generator.

Theorem 16 *The additive structure in the group $H^2(X, \mathbf{Z})$ corresponds with the tensor product in the family of one–dimensional complex vector bundles over X with respect to the isomorphism α.*

Proof.

The structure group of a one–dimensional complex vector bundle is $\mathbf{U}(1) = \mathbf{S}^1$. The tensor product is induced by the homomorphism

$$\varphi : \mathbf{S}^1 \times \mathbf{S}^1 \longrightarrow \mathbf{S}^1, \qquad (2.19)$$

which is defined by the formula

$$\varphi(z_1, z_2) = z_1 z_2.$$

The homomorphism (2.19) induces a mapping of the classifying spaces

$$\psi : \mathbf{CP}^\infty \times \mathbf{CP}^\infty \longrightarrow \mathbf{CP}^\infty \qquad (2.20)$$

such that universal bundle ξ over $\mathbf{CP}^\infty$ goes to tensor product

$$\psi^*(\xi) = \xi_1 \otimes \xi_2.$$

Hence the tensor product structure in the **Line** (X) is induced by the composition

$$X \xrightarrow{\Delta} X \times X \xrightarrow{f_1 \times f_2} \mathbf{CP}^\infty \times \mathbf{CP}^\infty \xrightarrow{\psi} \mathbf{CP}^\infty. \qquad (2.21)$$

On the other hand the mapping (2.21) induces the additive structure on the two–dimensional cohomology group. Indeed, the group structure (2.19) has the property that

$$\varphi(z, 1) = \varphi(1, z) = z.$$

Hence the same holds for (2.20):

$$\psi(x, x_0) = \psi(x_0, x) = x.$$

Therefore,

$$\psi^*(a) = (a \otimes 1) + (1 \otimes a) \in H^2(\mathbf{CP}^\infty, \mathbf{Z}).$$

Thus

$$
\begin{aligned}
\alpha([f_1] \otimes [f_2]) &= \\
&= (\psi(f_1 \times f_2)\Delta)^*(a) = (\Delta)^*(f_1 \times f_2)^*\psi^*(a) = \\
&= (\Delta)^*(f_1 \times f_2)^*(a \otimes 1 + 1 \otimes a) = \\
&= (\Delta)^*(f_1^*(a) \otimes 1 + 1 \otimes f_2^*(a)) = f_1^*(a) + f_2^*(a) = \\
&= \alpha([f_1]) + \alpha([f_2]).
\end{aligned}
$$

2.3 CONSTRUCTIONS OF THE CLASSIFYING SPACES

In this section we shall construct a universal principal G–bundle

$$p_G : E_G \longrightarrow B_G.$$

There are at least two different geometric constructions of a universal G–bundle. The problem is to construct a contractible space E_G with a free action of the group G.

2.3.1 Lie groups

Let us start with $G = \mathbf{U}(n)$. Consider the group $\mathbf{U}(N+n)$. The two subgroups:

$$\mathbf{U}(n) \subset \mathbf{U}(N + n)$$

and

$$\mathbf{U}(N) \subset \mathbf{U}(N + n)$$

are included as matrices

$$\left\| \begin{matrix} X & 0 \\ 0 & 1 \end{matrix} \right\|, \ X \in \mathbf{U}(n) \quad \text{and} \quad \left\| \begin{matrix} 1 & 0 \\ 0 & Y \end{matrix} \right\|, \ Y \in \mathbf{U}(N)$$

These subgroups commute and hence the group $\mathbf{U}(n)$ acts on the quotient space $\mathbf{U}(N + n)/\mathbf{U}(N)$. This action is free and the corresponding quotient space is homeomorphic to $\mathbf{U}(N + n)/\mathbf{U}(N) \oplus \mathbf{U}(n)$. Thus one has a principal $\mathbf{U}(n)$–bundle

$$\mathbf{U}(N + n)/\mathbf{U}(N) \longrightarrow \mathbf{U}(N + n)/\mathbf{U}(N) \oplus \mathbf{U}(n).$$

It is easy to see that total space of this bundle is homeomorphic to the complex Stieffel manifold $\mathbf{V}_{N+n,n}$ and the base is homeomorphic to complex Grassmann manifold $\mathbf{G}_{N+n,n}$. Now consider the inclusion

$$j : \mathbf{U}(N + n) \longrightarrow \mathbf{U}(N + n + 1)$$

defined by

$$T \longrightarrow \left\| \begin{matrix} T & 0 \\ 0 & 1 \end{matrix} \right\|, \ T \in \mathbf{U}(N + n).$$

The subgroup $\mathbf{U}(N)$ maps to the subgroup $\mathbf{U}(N+1)$:

$$\left\| \begin{matrix} 1 & 0 \\ 0 & Y \end{matrix} \right\| \longrightarrow \left\| \begin{matrix} \left\| \begin{matrix} 1 & 0 \\ 0 & Y \end{matrix} \right\| & 0 \\ 0 \quad 0 & 1 \end{matrix} \right\| = \left\| \begin{matrix} 1 & 0 & 0 \\ 0 & \left\| Y \right\| & 0 \\ 0 & 0 & 1 \end{matrix} \right\|, \quad Y \in \mathbf{U}(N),$$

and the subgroup $\mathbf{U}(n)$ maps to itself. Hence the inclusion j generates an inclusions of principal $\mathbf{U}(n)$–bundles

$$
\begin{array}{ccc}
\mathbf{U}(N+n)/\mathbf{U}(N) & \xrightarrow{\;j\;} & \mathbf{U}(N+n+1)/\mathbf{U}(N+1) \\
\downarrow p & & \downarrow p \\
\mathbf{U}(N+n)/(\mathbf{U}(N) \oplus \mathbf{U}(n)) & \xrightarrow{\;j\;} & \mathbf{U}(N+n+1)/(\mathbf{U}(N+1) \oplus \mathbf{U}(n))
\end{array}
$$

or

$$
\begin{array}{ccc}
\mathbf{V}_{N+n,n} & \xrightarrow{\;j\;} & \mathbf{V}_{N+n+1} \\
\downarrow p & & \downarrow p \\
\mathbf{G}_{N+n,n} & \xrightarrow{\;j\;} & \mathbf{G}_{N+n+1,n}
\end{array} \tag{2.22}
$$

Let $\mathbf{V}_{\infty,n}$ denote the direct limit

$$\mathbf{V}_{\infty,n} = \lim \mathbf{V}_{N+n,n},$$

and $\mathbf{G}_{\infty,n}$ the direct limit

$$\mathbf{G}_{\infty,n} = \lim \mathbf{G}_{N+n,n}.$$

The commutative diagram (2.22) induces the mapping

$$p : \mathbf{V}_{\infty,n} \longrightarrow \mathbf{G}_{\infty,n}. \tag{2.23}$$

The mapping (2.23) gives a principal $\mathbf{U}(n)$–bundle. To prove this it is sufficient to show that p gives a locally trivial bundle. Notice that by the definition of a direct limit, a set $U \subset \mathbf{G}_{\infty,n}$ is open iff for any N the set $U \cap \mathbf{G}_{N+n,n}$ is open in the space $\mathbf{G}_{N+n,n}$. Hence charts for the bundle (2.23) can be constructed as a union

$$\Omega = \cup \Omega_N$$

where

$$\Omega_N \subset \mathbf{G}_{N+n,n}$$

is a open chart in $\mathbf{G}_{N+n,n}$ such that

$$\Omega_N \subset \Omega_{N+1}.$$

The existence of the sequence $\{\Omega_N\}$ follows from Theorem 3 chapter 1 and from the fact that any closed cellular subset has a sufficiently small neighborhood which contracts to the closed set. Let us show that

$$\pi_k(\mathbf{V}_{\infty,n}) = 0. \tag{2.24}$$

Consider the bundle
$$\mathbf{U}(N+n)\longrightarrow \mathbf{V}_{N+n,n}$$
with fiber $\mathbf{U}(N)$. The inclusion
$$j : \mathbf{U}(N)\longrightarrow\mathbf{U}(N+n)$$
induces an isomorphism of the homotopy groups for $k < 2N$ (see example 2.2 of the previous section). Hence from the exact homotopy sequence

$$\longrightarrow\pi_k(\mathbf{U}(N))\longrightarrow\pi_k(\mathbf{U}(N+n))\longrightarrow\pi_k(\mathbf{V}_{N+n,n})\longrightarrow$$
$$\longrightarrow\pi_{k-1}(\mathbf{U}(N))\longrightarrow\pi_{k-1}(\mathbf{U}(N+n))\longrightarrow\ldots$$

it follows that
$$\pi_k(\mathbf{V}_{N+n,n}) = 0$$
for $k < 2N$. Since
$$\pi_k(\mathbf{V}_{\infty,n}) = \lim \pi_k(\mathbf{V}_{N+n,n})$$
the equality (2.24) holds for any k. Thus we have shown that (2.23) is a universal principal $\mathbf{U}(n)$–bundle.

Now let
$$\mathbf{G}\subset\mathbf{U}(n)$$
be a closed subgroup. Since we already know that there is a universal principal $\mathbf{U}(n)$–bundle (2.23), the group $\mathbf{U}(n)$ freely acts on the total space $\mathbf{V}_{\infty,n}$. Hence the group $\mathbf{G}$ also acts freely on the same space $\mathbf{V}_{\infty,n}$.

Proposition 3 *The mapping*
$$p_G : \mathbf{V}_{\infty,n}\longrightarrow\mathbf{B_G} = \mathbf{V}_{\infty,n}/\mathbf{G}$$
gives a locally trivial principal $\mathbf{G}$–bundle.

Proof.

Notice that the quotient mapping
$$q : \mathbf{U}(n)\longrightarrow\mathbf{U}(n)/\mathbf{G}$$
gives a locally trivial $\mathbf{G}$–bundle. Since the bundle (2.23) is locally trivial, and for any point $x \in \mathbf{V}_{\infty,n}$ there is an equivariant open neighborhood $W \ni x$ such that
$$W \equiv \mathbf{U}(n) \times W_0, \ W_0 \subset \mathbf{G}_{\infty,n},$$

then the mapping

$$p_G|W : W = \mathbf{U}(n) \times W_0 \longrightarrow \mathbf{U}(n)/\mathbf{G} \times W_0$$

gives a locally trivial bundle and the set $\mathbf{U}(n)/\mathbf{G} \times W_0$ is an open set. Hence the mapping p_G gives a locally trivial bundle. ∎

Proposition 4 *For each finite dimensional matrix group* $\mathbf{G}$ *there is a universal principal* $\mathbf{G}$*-bundle.*

Proof.

Consider a subgroup

$$\mathbf{H} \subset \mathbf{G}$$

such that $\mathbf{H}$ is deformation retract of $\mathbf{G}$, so

$$\pi_k(\mathbf{H}) \equiv \pi_k(\mathbf{G}). \tag{2.25}$$

Consider the commutative diagram

$$\begin{array}{ccc} E_\mathbf{H} & \longrightarrow & E_\mathbf{G} \\ \downarrow & & \downarrow \\ B_\mathbf{H} & \longrightarrow & B_\mathbf{G} \end{array} \tag{2.26}$$

The corresponding homotopy sequences give the following diagram

$$\begin{array}{ccccccccc} \longrightarrow & \pi_k(\mathbf{H}) & \longrightarrow & \pi_k(E_\mathbf{H}) & \longrightarrow & \pi_k(B_\mathbf{H}) & \longrightarrow & \pi_{k-1}(\mathbf{H}) & \longrightarrow \\ & \downarrow & & \downarrow & & \downarrow & & \downarrow & \\ \longrightarrow & \pi_k(\mathbf{G}) & \longrightarrow & \pi_k(E_\mathbf{G}) & \longrightarrow & \pi_k(B_\mathbf{G}) & \longrightarrow & \pi_{k-1}(\mathbf{G}) & \longrightarrow \end{array}$$

$$\tag{2.27}$$

Since

$$\pi_k(E_\mathbf{H}) = \pi_k(E_\mathbf{G}) = 0$$

for all k, and because of (2.25),

$$\pi_k(B_\mathbf{H}) \equiv \pi_k(B_\mathbf{G}).$$

On the other hand if there is universal principal $\mathbf{H}$-bundle then we can define $E_\mathbf{G}$ as the total space of the associated bundle over $B_\mathbf{H}$ with a new fiber $\mathbf{G}$. Then from (2.26) and (2.27) we have that

$$\pi_k(E_\mathbf{G}) = 0.$$

This means that for the group $\mathbf{G}$ there is universal principal $\mathbf{G}$-bundle.

To finish the proof, it is sufficient to notice that any matrix group has a compact subgroup which is deformation retract of the group. ∎

2.3.2 Milnor construction

There is another construction (due to J.Milnor) which can be used for any topological group. The idea consists of trying to kill homotopy groups of the total space starting from a simple **G**–bundle. Let us start from the bundle with one point as base

$$\mathbf{G} \longrightarrow \text{pt}. \tag{2.28}$$

The total space of the (2.28) consists of a single fiber **G**. In general,

$$\pi_0(\mathbf{G}) \neq 0.$$

To kill π_0 we attach a system of paths to **G** which connect pairs of points of the group in such a way that the left translation on the group can be extended to the paths. It is convenient to consider two copies of the group **G** and the family of intervals which connect each point of the first copy to each point of the second. This space is called the *join* of two copies of the group **G** and is denoted by $\mathbf{G} * \mathbf{G}$. Each point of $\mathbf{G} * \mathbf{G}$ is defined by a family $(\alpha_0, g_0, \alpha_1, g_1)$ where $g_0, g_1 \in \mathbf{G}$ and

$$0 \leq \alpha_0, \alpha_1 \leq 1, \alpha_0 + \alpha_1 = 1$$

are barycentric coordinates of a point on the unit interval, with the convention that

$$(0, g_0, 1, g_1) = (0, g_0', 1, g_1), \qquad g_0, g_0', g_1 \in \mathbf{G}$$
$$(1, g_0, 0, g_1) = (1, g_0, 0, g_1'), \qquad g_0, g_1, g_1' \in \mathbf{G}$$

The action of **G** on $\mathbf{G} * \mathbf{G}$ is defined by

$$h\,(\alpha_0, g_0, \alpha_1, g_1) = (\alpha_0, hg_0, \alpha_1, hg_1),\ h \in \mathbf{G}.$$

Similarly, the join of $(k + 1)$ copies of the group **G** is denoted by

$$\mathbf{G}^{*k} = \mathbf{G} * \mathbf{G} * \cdots * \mathbf{G} \ \text{(k+1 times)}.$$

A point of the join has the form

$$x = (\alpha_0, g_0, \alpha_1, g_1, \dots, \alpha_k, g_k) \tag{2.29}$$

where

$$g_0, g_1, \dots g_k \in \mathbf{G}$$

are arbitrary elements and

$$(\alpha_0, \alpha_1, \ldots \alpha_k) : \alpha_i \geq 0, \; \alpha_0 + \alpha_1 + \cdots + \alpha_k = 1$$

are barycentric coordinates of the point in a k–dimensional simplex; this description (2.29) of the point x should be factorized by relation

$$(\alpha_0, g_0, \alpha_1, g_1, \ldots, \alpha_k, g_k) \equiv (\alpha'_0, g'_0, \alpha'_1, g'_1, \ldots, \alpha'_k, g'_k)$$

iff

$$\alpha_i = \alpha'_i \text{ for all } i$$

and

$$g_i = g'_i \text{ for each } i \text{ when } \alpha_i = 0.$$

The group $\mathbf{G}$ acts on the $\mathbf{G}^{*k}$ by formula

$$h(\alpha_0, g_0, \alpha_1, g_1, \ldots, \alpha_k, g_k) = (\alpha_0, hg_0, \alpha_1, hg_1, \ldots, \alpha_k, hg_k), \; h \in \mathbf{G}.$$

Proposition 5 *When* $\leq k - 1$,

$$\pi_j(\mathbf{G}^{*(k+1)}) = 0.$$

Proof.

The join $X * Y$ is the union

$$X * Y = ((CX \times Y)) \cup ((X \times CY)) \tag{2.30}$$

where CX, CY are cones over X and Y. The union (2.30) can represented as the union

$$X * Y = ((CX \times Y)) \cup CY) \cup ((X \times CY)) \cup CX) = P_1 \cup P_2$$

where

$$\begin{aligned} P_1 &= ((CX \times Y) \cup CY), \\ P_2 &= ((X \times CY) \cup CX). \end{aligned}$$

Both spaces P_1 and P_2 are contractible and intersection is

$$P_3 = P_1 \cap P_2 = (X \times Y) \cup CX \cup CY.$$

Hence both P_1 and P_2 are homotopy equivalent to the cone CP_3. So the union $P_1 \cup P_2$ is homotopy equivalent to the suspension SP_3. By induction, the space $\mathbf{G}^{*(k+1)}$ is homotopy equivalent to the $S^k Z$ for some Z. It is easy to check that

$$\pi_0(\mathbf{G}^{*(k+1)}) = H_0(\mathbf{G}^{*(k+1)}) = 0.$$

Hence, using the Hurewich theorem

$$\pi_j(\mathbf{G}^{*(k+1)}) = H_j(\mathbf{G}^{*(k+1)}) = 0 \text{ for } j \leq k.$$

Hence

$$\mathbf{G}_\infty = \lim \mathbf{G}^{*k}$$

has trivial homotopy groups. ∎

Examples

1. For any real vector bundle the structure group can be taken as $\mathbf{O}(n)$, for some n. The corresponding classifying space can be constructed as a limit of real Grassmann manifolds:

$$\mathbf{BO}(n) = \lim \mathbf{G}^R_{N+n,n} = \mathbf{G}^R_{\infty,n}$$

2. Let

$$\mathbf{EU}(n) \longrightarrow \mathbf{BU}(n)$$

be the universal principal $\mathbf{U}(n)$–bundle. Then

$$\mathbf{BO}(n) \equiv \mathbf{EU}(n)/\mathbf{O}(n),$$

and hence there is the quotient mapping

$$c : \mathbf{BO}(n) \longrightarrow \mathbf{BU}(n)$$

which gives a locally trivial bundle with the fiber $\mathbf{U}(n)/\mathbf{O}(n)$. The mapping c corresponds to the operation of complexification of real vector bundles which is also denoted by $c\xi$.

3. The family of all real vector bundles over a cellular space X, for which the complexifications are trivial bundles, is given by the image of the homomorphism

$$\mathbf{Im} \; ([X, \mathbf{U}(n)/\mathbf{O}(n)] \longrightarrow [X, \mathbf{BO}(n)]) . \tag{2.31}$$

4. The family $[X, \mathbf{U}(n)/\mathbf{O}(n)]$ is bigger than the image (2.31). It is easy to see that each continuous mapping

$$f : X \longrightarrow \mathbf{U}(n)/\mathbf{O}(n)$$

defines a real vector bundle ξ for which the complexification is trivial and gives concrete trivialization of the bundle $c\xi$.

Proposition 6 *Each element of the family $[X, \mathbf{U}(n)/\mathbf{O}(n)]$ defines a real vector ξ bundle with a homotopy class of trivializations of the bundle $c\xi$.*

Proof.

The mapping

$$\mathbf{U}(n) \longrightarrow \mathbf{U}(n)/\mathbf{O}(n)$$

gives a principal $\mathbf{O}(n)$–bundle. Hence each mapping

$$f : X \longrightarrow \mathbf{U}(n)/\mathbf{O}(n)$$

induces a principal $\mathbf{O}(n)$– bundle over X and an equivariant mapping of total spaces

$$
\begin{array}{ccc}
\xi_O & \xrightarrow{g} & \mathbf{U}(n) \\
\downarrow & & \downarrow \\
X & \xrightarrow{f} & \mathbf{U}(n)/\mathbf{O}(n)
\end{array}
.
$$

Let ξ_U be the total space of the principal $\mathbf{U}(n)$–bundle associated with the complexification of ξ_O. Then the space ξ_O lies in the space ξ_U and this inclusion $\xi_O \subset \xi_U$ is equivariant with respect to the actions of the group $\mathbf{O}(n)$ on ξ_O and the group $\mathbf{U}(n)$ on ξ_U. Therefore, the map g extends uniquely to an equivariant map

$$\bar{g} : \xi_U \longrightarrow \mathbf{U}(n).$$

The latter defines a trivialization of the bundle ξ_U. If

$$f_1, f_2 : X \longrightarrow \mathbf{U}(n)/\mathbf{O}(n)$$

are homotopic then there is a bundle ξ_O over $X \times I$ with a trivialization of the complexification. Since the restrictions of the bundle ξ_O to the $X \times \{t\}$ are all isomorphic, the trivializations of $\xi_U|_{X \times \{0\}}$ and $\xi_U|_{X \times \{1\}}$ are homotopic. Conversely, let ξ_O be a principal $\mathbf{O}(n)$–bundle and fix a trivialization of the complexification ξ_U. This means that there is map $\bar{g}$ giving a commutative diagram

$$
\begin{array}{ccccc}
\xi_U & \xrightarrow{\bar{g}} & \mathbf{U}(n) & \subset & \mathbf{EU}(n) \\
\downarrow & & \downarrow & & \downarrow \\
X & \longrightarrow & \{pt\} & \subset & \mathbf{BU}(n)
\end{array}
$$

Hence there is a commutative diagram

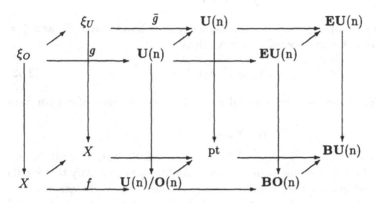

The mapping

$$f : X \longrightarrow \mathbf{U}(n)/\mathbf{O}(n)$$

is induced by the mapping $\bar{g}$ and generates the bundle ξ_O and the trivialization $\bar{g}$ of the bundle $c\xi_O$. Hence the family of homotopy classes of mappings $[X, \mathbf{U}(n)/\mathbf{O}(n)]$ can be interpreted as the family of the equivalence classes of pairs (ξ, φ) where ξ is real bundle over X and φ is an isomorphism of $c\xi$ with a trivial bundle. Two pairs (ξ_1, φ_1) and (ξ_2, φ_2) are considered equivalent if the bundles ξ_1 and ξ_2 are isomorphic and the isomorphisms φ_1 and φ_2 are homotopic in the class of isomorphisms.

2.4 CHARACTERISTIC CLASSES

In previous sections we showed that, generally speaking, any bundle can be obtained as an inverse image or pull back of a universal bundle by a continuous mapping of the base spaces. In particular, isomorphisms of vector bundles over X are characterized by homotopy classes of continuous mappings of the space X to the classifying space $\mathbf{BO}(n)$ (or $\mathbf{BU}(n)$ for complex bundles). But it is usually difficult to describe homotopy classes of maps from X into $\mathbf{BO}(n)$. Instead, it is usual to study certain invariants of vector bundles defined in terms of the homology or cohomology groups of the space X.

Following this idea, we use the term *characteristic class* for a correspondence α which associates to each n-dimensional vector bundle ξ over X a cohomology class $\alpha(\xi) \in H^*(X)$ with some fixed coefficient group for the cohomology groups. In addition, we require functoriality : if

$$f : X \longrightarrow Y$$

is a continuous mapping, η an n-dimensional vector bundle over Y, and $\xi = f^*(\eta)$ the pull-back vector bundle over X, then

$$\alpha(\xi) = f^*(\alpha(\eta)), \tag{2.32}$$

where in (2.32) f^* denotes the induced natural homomorphism of cohomology groups

$$f^* : H^*(Y) \longrightarrow H^*(X).$$

If we know the cohomology groups of the space X and the values of all characteristic classes for given vector bundle ξ, then might hope to identify the bundle ξ, that is, to distinguish it from other vector bundles over X. In general, this hope is not justified. Nevertheless, the use of characteristic classes is a standard technique in topology and in many cases gives definitive results.

Let us pass on to study properties of characteristic classes.

Theorem 17 *The family of all characteristic classes of n-dimensional real (complex) vector bundles is in one-to-one correspondence with the cohomology ring $H^*(\mathbf{BO}(n))$ (respectively, with $H^*(\mathbf{BU}(n)))$.*

Proof.

Let ξ_n be the universal bundle over the classifying space $\mathbf{BO}(n)$ and α a characteristic class. Then $\alpha(\xi_n) \in H^*(\mathbf{BO}(n))$ is the associated cohomology class. Conversely, if $x \in H^*(\mathbf{BO}(n))$ is arbitrary cohomology class then a characteristic class α is defined by the following rule: if $f : X \longrightarrow \mathbf{BO}(n)$ is continuous map and $\xi = f^*(\xi_n)$ put

$$\alpha(\xi) = f^*(x) \in H^*(X). \tag{2.33}$$

Let us check that the correspondence (2.33) gives a characteristic class. If

$$g : X \longrightarrow Y$$

is continuous map and

$$h : Y \longrightarrow \mathbf{BO}(n)$$

is a map such that

$$\eta = g^*(\xi_n), \ \xi = g^*(\eta),$$

then

$$\alpha(\xi) = \alpha((hg)^*(\xi_n)) = (hg)^*(x) = g^*(h^*(x)) = g^*(\alpha(h^*(\xi_n))) = g^*(\alpha(\eta)).$$

If

$$f : \mathbf{BO}(n) \longrightarrow \mathbf{BO}(n)$$

is the identity mapping then

$$\alpha(\xi_n) = f^*(x) = x.$$

Hence the class α corresponds to the cohomology class x.   ∎

We now understand how characteristic classes are defined on the family of vector bundles of a fixed dimension. The characteristic classes on the family of all vector bundles of any dimension should be as follows: a characteristic class is a sequence

$$\alpha = \{\alpha_1, \alpha_2, \ldots, \alpha_n, \ldots\} \tag{2.34}$$

where each term α_n is a characteristic class defined on vector bundles of dimension n.

Definition 5 *A class α of the form (2.34) is said to be stable if the following condition holds:*

$$\alpha_{n+1}(\xi \oplus 1) = \alpha_n(\xi), \tag{2.35}$$

for any n–dimensional vector bundle ξ.

In accordance with Theorem 17 one can think of α_n as a cohomology class

$$\alpha_n \in H^*(\mathbf{BO}(n)).$$

Let

$$\varphi : \mathbf{BO}(n) \longrightarrow \mathbf{BO}(n+1)$$

be the natural mapping for which

$$\varphi^*(\xi_{n+1}) = \xi_n \oplus 1.$$

This mapping φ is induced by the natural inclusion of groups

$$\mathbf{O}(n) \subset \mathbf{O}(n+1).$$

Then the condition (2.35) is equivalent to:

$$\varphi^*(\alpha_{n+1}) = \alpha_n. \tag{2.36}$$

Consider the sequence

$$\mathbf{BO}(1) \longrightarrow \mathbf{BO}(2) \longrightarrow \ldots \longrightarrow \mathbf{BO}(n) \longrightarrow \mathbf{BO}(n+1) \longrightarrow \ldots$$

and the direct limit

$$\mathbf{BO} = \lim_{\longrightarrow} \mathbf{BO}(n).$$

Let

$$H^*(\mathbf{BO}) = \lim_{\longleftarrow} H^*(\mathbf{BO}(n)).$$

Condition 2.36 means that the family of stable characteristic classes is in one-to-one correspondence with the cohomology ring $H^*(\mathbf{BO})$.

Now we consider the case of cohomology with integer coefficients.

Theorem 18 *The ring $H^*(\mathbf{BU}(n); \mathbf{Z})$ of integer cohomology classes is isomorphic to the polynomial ring $\mathbf{Z}[c_1, c_2, \ldots, c_n]$, where*

$$c_k \in H^{2k}(\mathbf{BU}(n); \mathbf{Z}).$$

The generators $\{c_1, c_2, \ldots, c_n\}$ can be chosen such that

1. *the natural mapping*

$$\varphi : \mathbf{BU}(n) \longrightarrow \mathbf{BU}(n+1) \tag{2.37}$$

   satisfies

$$\begin{aligned}
\varphi^*(c_k) &= c_k, \ k = 1, 2, \ldots, n, \\
\varphi^*(c_{n+1}) &= 0;
\end{aligned} \tag{2.38}$$

2. *for a direct sum of vector bundles we have the relations*

$$c_k(\xi \oplus \eta) = c_k(\xi) + c_{k-1}(\xi)c_1(\eta) +$$
$$+ \quad c_{k-2}(\xi)c_2(\eta) + \ldots + c_1(\xi)c_{k-1}(\eta) + c_k(\eta) =$$
$$= \sum_{\alpha+\beta=k} c_\alpha(\xi)c_\beta(\eta), \tag{2.39}$$

where $c_0(\xi) = 1$.

The condition (2.38) means that the sequence

$$\{0, \ldots, c_k, c_k, \ldots, c_k, \ldots\}$$

is a stable characteristic class which will also be denoted by c_k. This notation was used in (2.39). If $\dim \xi < k$ then by (2.38) it follows that $c_k(\xi) = 0$.

Formula (2.39) can be written in a simpler way. Put

$$c = 1 + c_1 + c_2 + \ldots + c_k + \ldots \tag{2.40}$$

The formal series (2.40) has a well defined value on any vector bundle ξ since in the infinite sum (2.40) only a finite number of the summands will be nonzero:

$$c(\xi) = 1 + c_1(\xi) + c_2(\xi) + \ldots + c_k(\xi), \text{ if } \dim \xi = k.$$

Hence from (2.39) we see that

$$c(\xi \oplus \eta) = c(\xi)c(\eta). \tag{2.41}$$

Conversely, the relations (2.39) may be obtained from (2.41) by considering the homogeneous components in (2.41).

Proof.

Let us pass now to the proof of Theorem 18. The method we use for the calculation of cohomology groups of the space $\mathbf{BU}(n)$ involves spectral sequences for bundles. Firstly, using spectral sequences, we calculate the cohomology groups of unitary group $\mathbf{U}(n)$.

Since

$$H^0(\mathbf{S}^n) = \mathbf{Z},$$
$$H^n(\mathbf{S}^n) = \mathbf{Z},$$
$$H^k(\mathbf{S}^n) = 0, \text{ when } k \neq 0 \text{ and } k \neq n,$$

the cohomology ring

$$H^*(\mathbf{S}^n) =$$
$$= \oplus_k H^k(\mathbf{S}^n) =$$
$$= H^0(\mathbf{S}^n) \oplus H^n(\mathbf{S}^n)$$

is a free exterior algebra over the ring of integers $\mathbf{Z}$ with a generator $a_n \in H^n(\mathbf{S}^n)$. The choice of the generator a_n is not unique: one can change a_n for $(-a_n)$. We write

$$H^*(\mathbf{S}^n) = \Lambda(a_n).$$

Now consider the bundle $\mathbf{U}(2) \longrightarrow \mathbf{S}^3$ with fibre $\mathbf{S}^1$. The second term of spectral sequence for this bundle is

$$E_2^{*,*} = \sum_{p,q} E_2^{p,q} =$$
$$= H^*(\mathbf{S}^3, H^*(\mathbf{S}^1)) = H^*(\mathbf{S}^3) \otimes H^*(\mathbf{S}^1) =$$
$$= \Lambda(a_3) \otimes \Lambda(a_1) = \Lambda(a_1, a_3).$$

The differential d_2 vanishes except possibly on the generator

$$1 \otimes a_1 \in E_2^{0,1} = H^0\left(\mathbf{S}^3, H^1(\mathbf{S}^1)\right).$$

But then

$$d_2(1 \otimes a_1) \in E_2^{2,0} = H^2\left(\mathbf{S}^3, H^0(\mathbf{S}^1)\right) = 0.$$

Hence

$$d_2(1 \otimes a_1) = 0,$$
$$d_2(a_3 \otimes 1) = 0,$$
$$d_2(a_3 \otimes a_1) = d_2(a_3 \otimes 1)a_1 - a_3 d_2(1 \otimes a_1) = 0.$$

Hence d_2 is trivial and therefore

$$E_3^{p,q} = E_2^{p,q}.$$

Similarly $d_3 = 0$ and

$$E_4^{p,q} = E_3^{p,q} = E_2^{p,q}.$$

Continuing, $d_n = 0$ and

$$E_{n+1}^{*,*} = E_n^{*,*} = \ldots = E_2^{*,*} = \Lambda(a_1, a_3),$$
$$E_\infty^{*,*} = \Lambda(a_1, a_3).$$

The cohomology ring $H^*(\mathbf{U}(2))$ is associated to the ring $\Lambda(a_1, a_3)$, that is, the ring $H^*(\mathbf{U}(2))$ has a filtration for which the resulting factors are isomorphic to the homogeneous summands of the ring $\Lambda(a_1, a_3)$. In each dimension, $n = p+q$, the groups $E_\infty^{p,q}$ vanish except for a single value of p, q. Hence

$$
\begin{aligned}
H^0(\mathbf{U}(2)) &= \mathbf{Z}, \\
H^1(\mathbf{U}(2)) &= E_\infty^{1,0}, \\
H^3(\mathbf{U}(2)) &= E_\infty^{0,3}, \\
H^4(\mathbf{U}(2)) &= E_\infty^{1,3}.
\end{aligned}
$$

Let

$$u_1 \in H^1(\mathbf{U}(2)), \ u_3 \in H^3(\mathbf{U}(2))$$

be generators which correspond to a_1 and a_3, respectively. As $a_1 a_3$ is a generator of the group $E_\infty^{1,3}$, the element $u_1 u_3$ is a generator of the group $H^4(\mathbf{U}(2))$.

It is useful to illustrate our calculation as in figure 2.1, where the nonempty cells show the positions of the generators the groups $E_s^{p,q}$ for each fixed s–level of the spectral sequence.

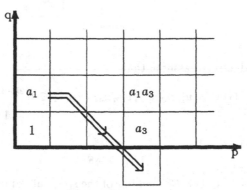

Figure 2.1

For brevity the tensor product sign $\otimes$ is omitted. The arrow denotes the action of the differential d_s for $s = 2, 3$. Empty cells denote trivial groups.

Thus we have shown

$$H^*(\mathbf{U}(2)) = \Lambda(u_1, u_3).$$

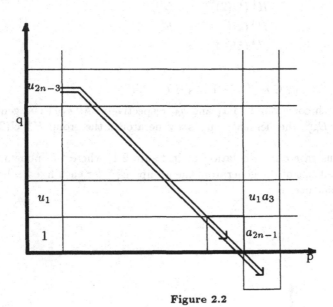

Figure 2.2

Proceeding inductively, assume that

$$H^* \left(\mathbf{U}(n-1) \right) = \Lambda(u_1, u_3, \ldots, u_{2n-3}), \ u_{2k-1} \in H^{2k-1} \left(\mathbf{U}(n-1) \right),$$
$$1 \leq k \leq n-1,$$

and consider the bundle

$$\mathbf{U}(n) \longrightarrow \mathbf{S}^{2n-1}$$

with fiber $\mathbf{U}(n-1)$. The E_2 member of the spectral sequence has the following form:

$$E_2^{*,*} = H^* \left(\mathbf{S}^{2n-1}; H^* \left(\mathbf{U}(n-1) \right) \right) =$$
$$= \Lambda(a_{2n-1}) \otimes \Lambda(u_1, \ldots, u_{2n-3}) = \Lambda(u_1, u_3, \ldots, u_{2n-3}, a_{2n-1}).$$

(see fig.2.2).

The first possible nontrivial differential is d_{2n-1}. But

$$d_{2n-1}(u_k) = 0, \ k = 1, 3, \ldots, 2n - 3,$$

and each element $x \in E^{0,q}$ decomposes into a product of the elements u_k. Thus $d_{2n-1}(x) = 0$. Similarly, all subsequent differentials d_s are trivial. Thus

$$E_\infty^{p,q} = \ldots = E_s^{p,q} = \ldots = E_2^{p,q} =$$
$$= H^p \left(\mathbf{S}^{2n-1}, H^q \left(\mathbf{U}(n-1) \right) \right),$$
$$E_\infty^{*,*} = \Lambda(u_1, u_3, \ldots, u_{2n-3}, u_{2n-1}).$$

Let now show that the ring $H^* \left(\mathbf{U}(n) \right)$ is isomorphic to exterior algebra

$$\Lambda(u_1, u_3, \ldots, u_{2n-3}, u_{2n-1}).$$

Since the group $E_\infty^{*,*}$ has no torsion, there are elements $v_1, v_3, \ldots, v_{2n-3} \in H^* \left(\mathbf{U}(n) \right)$ which go to $u_1, u_3, \ldots, u_{2n-3}$ under the inclusion $\mathbf{U}(n-1) \subset \mathbf{U}(n)$. All the v_k are odd dimensional. Hence elements of the form $v_1^{\varepsilon_1} v_3^{\varepsilon_3} \cdots v_{2n-3}^{\varepsilon_{2n-3}}$ where $\varepsilon_k = 0, 1$ generate a subgroup in the group $H^* \left(\mathbf{U}(n) \right)$ mapping isomorphically onto the group $H^* \left(\mathbf{U}(n-1) \right)$. The element $a_{2n-1} \in E_\infty^{2n-1,0}$ has filtration $2n - 1$. Hence the elements of the form

$$v_1^{\varepsilon_1} v_3^{\varepsilon_3} \cdots v_{2n-3}^{\varepsilon_{2n-3}} a_{2n-1}$$

form a basis of the group $E_\infty^{2n-1,0}$. Thus the group $H^* \left(\mathbf{U}(n) \right)$ has a basis consisting of the elements

$$v_1^{\varepsilon_1} v_3^{\varepsilon_3} \cdots v_{2n-3}^{\varepsilon_{2n-3}} a_{2n-1}^{\varepsilon_{2n-1}}, \ \varepsilon_k = 0, 1.$$

Thus

$$H^* \left(\mathbf{U}(n) \right) = \Lambda = (v_1, v_3, \ldots, v_{2n-3}, v_{2n-1}).$$

∎

Now consider the bundle

$$\mathbf{EU}(1) \longrightarrow \mathbf{BU}(1)$$

with the fibre $\mathbf{U}(1) = \mathbf{S}^1$. From the exact homotopy sequence

$$\pi_1 \left(\mathbf{EU}(1) \right) \longrightarrow \pi_1 \left(\mathbf{BU}(1) \right) \longrightarrow \pi_0 \left(\mathbf{S}^1 \right)$$

it follows that

$$\pi_1 \left(\mathbf{BU}(1) \right) = \mathbf{0}.$$

At this stage we do not know the cohomology of the base, but we know the cohomology of the fibre

$$H^* \left(\mathbf{S}^1 \right) = \Lambda(u_1),$$

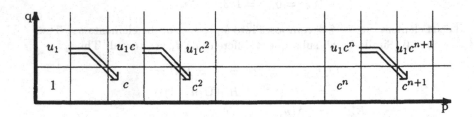

Figure 2.3

and cohomology of the total space

$$H^* \left(\mathbf{EU}(1) \right) = 0.$$

This means that

$$E_\infty^{p,q} = \cap_s E_s^{p,q} = 0.$$

We know that

$$E_2^{p,q} = H^p \left(\mathbf{BU}(1) \right), H^q(\mathbf{S}^1),$$

and hence

$$E_2^{p,q} = 0 \text{ when } q \geq 2.$$

In the figure 2.3 nontrivial groups can only occur in the two rows with $q = 0$ and $q = 1$. Moreover,

$$E_2^{p,1} \sim E_2^{p,0} \otimes u_1 \sim E_2^{p,0}.$$

But

$$E_2^{p,q} = 0 \text{ for } q \geq 2,$$

and it follows that

$$E_s^{p,q} = 0 \text{ for } q \geq 2.$$

Hence all differentials from d_3 on are trivial and so

$$E_3^{p,q} = \ldots = E_\infty^{p,q} = 0.$$

Also

$$E_3^{p,q} = H(E_2^{p,q}, d_2)$$

and thus the differential

$$d_2 : E_2^{p,1} \longrightarrow E^{p+2,0}$$

is an isomorphism. Putting

$$c = d_2(u_1),$$

we have

$$d_2(u_1 c^k) = d_2(u_1)c^k = c^{k+1}.$$

Hence the cohomology ring of the space $\mathbf{BU}(1)$ is isomorphic to the polynomial ring with a generator c of the dimension 2:

$$H^*(\mathbf{BU}(1)) = \mathbf{Z}[c].$$

Now assume that

$$H^*(\mathbf{BU}(n-1)) = \mathbf{Z}[c_1, \ldots, c_{n-1}]$$

and consider the bundle

$$\mathbf{BU}(n-1) \longrightarrow \mathbf{BU}(n)$$

with the fiber $\mathbf{U}(n)/\mathbf{U}(n-1) = \mathbf{S}^{2n-1}$. The exact homotopy sequence gives us that

$$\pi_1(\mathbf{BU}(n)) = 0.$$

We know the cohomology of the fiber $\mathbf{S}^{2n-1}$ and the cohomology of the total space $\mathbf{BU}(n-1)$. The cohomology of the latter is not trivial but equals the ring $\mathbf{Z}[c_1, \ldots, c_{n-1}]$. Therefore, in the spectral sequence only the terms $E_s^{p,0}$ and $E_s^{p,2n-1}$ may be nontrivial and

$$E_2^{p,2n-1} = E_2^{p,0} \otimes a_{2n-1} = H^p(\mathbf{BU}(n)) \otimes = a_{2n-1}.$$

(see figure 2.4).

Hence the only possible nontrivial differential is d_{2n} and therefore

$$E_2^{p,q} = \ldots = E_{2n}^{p,q},$$

$$H(E_{2n}^{p,q}, d_{2n}) = E_{2n+1}^{p,q} = \ldots = E_\infty^{p,q}$$

It is clear that if $n = p + q$ is odd then $E_\infty^{p,q} = 0$. Hence the differential

$$d_{2n} : E_{2n}^{0,2n-1} \longrightarrow E_{2n}^{2n,0}$$

is a monomorphism. If $p + q = k < 2n - 1$ then $E_2^{p,q} = E_\infty^{p,q}$. Hence for odd $k \leq 2n - 1$, the groups $E_{2n}^{k,0}$ are trivial. Hence the differential

$$d_{2n} : E_{2n}^{k,2n-1} \longrightarrow E_{2n}^{k+2n,0}$$

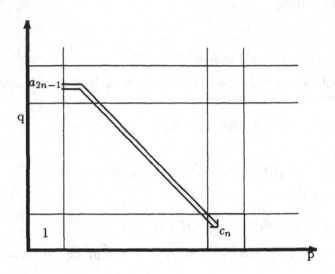

Figure 2.4

is a monomorphism for $k \le 2n$. This differential makes some changes in the term $E_{2n+1}^{k+2n,0}$ only for even $k \le 2n$. Hence, for odd $k \le 2n$, we have

$$E_{2n+1}^{k+2n,0} = 0.$$

Thus the differential
$$d_{2n} : E_{2n}^{k,2n-1} \longrightarrow E_{2n}^{k+2n,0}$$
is a monomorphism for $k \le 4n$. By induction one can show that

$$E_{2n}^{k,0} = 0$$

for arbitrary odd k, and the differential

$$d_{2n} : E_{2n}^{k,2n-1} \longrightarrow E_{2n}^{k+2n,0}$$

is a monomorphism. Hence

$$E_{2n+1}^{k,2n-1} = 0$$
$$E_{2n+1}^{k+2n,0} = E_{2n}^{k2n,0}/E_{2n}^{k,2n-1} = E_{\infty}^{k+2n,0}.$$

The ring $H^*\left(\mathbf{BU}(n-1)\right)$ has no torsion and in the term $E_{\infty}^{p,q}$ only one row is nontrivial ($q = 0$). Hence the groups $E_{\infty}^{k+2n,0}$ have no torsion. This means that image of the differential d_{2n} is a direct summand.

Let
$$c_n = d_{2n}(a_{2n-1}).$$
and then
$$d_{2n}(x a_{2n-1}) = x c_{2n}.$$
It follows that the mapping
$$H^*(\mathbf{BU}(n)) \longrightarrow H^*(\mathbf{BU}(n))$$
defined by the formula
$$x \longrightarrow c_n x$$
is a monomorphism onto a direct summand and the quotient ring is isomorphic to the ring $H^*(\mathbf{BU}(n-1)) = \mathbf{Z}[c_1, \ldots, c_{n-1}]$. Thus
$$H^*(\mathbf{BU}(n)) = \mathbf{Z}[c_1, \ldots, c_{n-1}, c_n].$$

Consider now the subgroup
$$\mathbf{T}^n = \mathbf{U}(1) \times \ldots \times \mathbf{U}(1) \ (n \text{ times }) \subset \mathbf{U}(n) \tag{2.42}$$
of diagonal matrices. The natural inclusion $\mathbf{T}^n \subset \mathbf{U}(n)$ induces a mapping
$$j_n : \mathbf{BT}^n \longrightarrow \mathbf{BU}(n). \tag{2.43}$$
But
$$\mathbf{BT}^n = \mathbf{BU}(1) \times \ldots \times \mathbf{BU}(1),$$
and hence
$$H^*(\mathbf{BT}^n) = \mathbf{Z}[t_1, \ldots, t_n].$$

Lemma 4 *The homomorphism*
$$j_n^* : \mathbf{Z}[c_1, \ldots, c_n] \longrightarrow \mathbf{Z}[t_1, \ldots, t_n]$$
induced by the mapping (2.43) is a monomorphism onto the direct summand of all symmetric polynomials in the variables $(t_1, \ldots, t_n)$.

Proof.

Let
$$\alpha : \mathbf{U}(n) \longrightarrow \mathbf{U}(n)$$

be the inner automorphism of the group induced by permutation of the basis of the vector space on which the group $U(n)$ acts. The automorphism α acts on diagonal matrices by permutation of the diagonal elements. In other words, α permutes the factors in the group (2.42). The same is true for the classifying spaces and the following diagram

$$
\begin{array}{ccc}
\mathbf{BT}^n & \xrightarrow{j_n} & \mathbf{BU}(n) \\
\downarrow{\scriptstyle\alpha} & & \downarrow{\scriptstyle\alpha} \\
\mathbf{BT}^n & \xrightarrow{j_n} & \mathbf{BU}(n)
\end{array}
$$

is commutative. The inner automorphism α is homotopic to the identity since the group $U(n)$ is connected. Hence, on the level of cohomology, the following diagram

$$
\begin{array}{ccc}
H^*(\mathbf{BT}^n) & \xleftarrow{j_n^*} & H^*(\mathbf{BU}(n)) \\
\uparrow{\scriptstyle\alpha^*} & & \uparrow{\scriptstyle\alpha^*} \\
H^*(\mathbf{BT}^n) & \xleftarrow{j_n^*} & H^*(\mathbf{BU}(n))
\end{array}
$$

or

$$
\begin{array}{ccc}
\mathbf{Z}[c_1,\ldots,c_n] & \xrightarrow{j_n^*} & \mathbf{Z}[t_1,\ldots,t_n] \\
\downarrow{\scriptstyle\alpha^*} & & \downarrow{\scriptstyle\alpha^*} \\
\mathbf{Z}[c_1,\ldots,c_n] & \xrightarrow{j_n^*} & \mathbf{Z}[t_1,\ldots,t_n]
\end{array}.
$$

is commutative. The left homomorphism α^* is the identity, whereas the right permutes the variables $(t_1,\ldots,t_n)$. Hence, the image of j_n^* consists of symmetric polynomials.

Now let us prove that the image of j_n^* is a direct summand. For this it is sufficient to show that the inclusion

$$
U(k) \times U(1) \subset U(k+1)
$$

induces a monomorphism in cohomology onto a direct summand. Consider the corresponding bundle

$$
\mathbf{B}\left(U(k) \times U(1)\right) \longrightarrow \mathbf{BU}(k+1)
$$

with fiber

$$
U(k+1)/\left(U(k) \times U(1)\right) = \mathbf{CP}^k.
$$

The E_2 term of the spectral sequence is

$$
E_2^{*,*} = H^*\left(\mathbf{BU}(k+1); H^*(\mathbf{CP}^k)\right). \tag{2.44}
$$

For us it is important here that the only terms of (2.44) which are nontrivial occur when p and q are even. Hence all differentials d_s are trivial and

$$
E_2^{p,q} = E_\infty^{p,q}.
$$

None of these groups have any torsion. Hence the group

$$H^p\left(\mathbf{BU}(k+1);\mathbf{Z}\right) = E_\infty^{p,0} \subset H^*\left(\mathbf{BU}(k) \times \mathbf{BU}(1)\right)$$

is a direct summand.

It is very easy to check that the rank of the group $H^k\left(\mathbf{BU}(n)\right)$ and the subgroup of symmetric polynomials of the degree k of variables $(t_1, \ldots, t_n)$ are the same.

∎

Using Lemma 4, choose generators

$$c_1, \ldots, c_n \in H^*\left(\mathbf{BU}(n)\right)$$

as inverse images of the elementary symmetric polynomials in the variables

$$t_1, \ldots, t_n \in H^*\left(\mathbf{BU}(1) \times \ldots \times \mathbf{BU}(1)\right).$$

Then the condition (2.38) follows from the fact that the element c_{n+1} is mapped by j_{n+1}^* to the product $t_1 \cdot \ldots \cdot t_{n+1}$, which in turn is mapped by the inclusion (2.37) to zero. Condition 2.39 follows from the properties of the elementary symmetric polynomials:

$$\sigma_k(t_1, \ldots, t_n, t_{n+1}, \ldots, t_{n+m}) = \sum_{\alpha+\beta=k} \sigma_\alpha(t_1, \ldots, t_n)\sigma_\beta(t_{n+1}, \ldots, t_{n+m}).$$

The proof of Theorem 18 is finished. ∎

The generators $c_1, \ldots, c_n$ are not unique but only defined up to a choice of signs for the generators $t_1, \ldots, t_n$. Usually the sign of the t_k is chosen in such way that for the Hopf bundle over $\mathbf{CP}^1$ the value of c_1 on the fundamental circle is equal to 1.

The characteristic classes c_k are called *Chern classes*. If X is complex analytic manifold then characteristic classes of the tangent bundle TX are simply called *characteristic classes of manifold* and one writes

$$c_k(X) = c_k(TX).$$

Example

Consider the complex projective space $\mathbf{CP}^n$. It was shown in the section 6 of the chapter 1 that the tangent bundle $T\mathbf{CP}^n$ satisfies the relation

$$T\mathbf{CP}^n \oplus \bar{1} = (n+1)\xi^*$$

where ξ is the Hopf bundle. Hence

$$c_k\left(T\mathbf{CP}^n\right) = c_k\left(T\mathbf{CP}^n \oplus \bar{1}\right) = c_k\left((n+1)\xi^*\right) =$$

$$= \sigma_k(-t_1, -t_2, \ldots, -t_n)|_{t_k=t} = (-1)^k \binom{n+1}{k} t^k.$$

From this we can see that the tangent bundle of $\mathbf{CP}^n$ is nontrivial.

We have studied in detail the algebraic structure of characteristic classes of complex bundles. The proof was based on simple geometric properties of the structure group $\mathbf{U}(n)$ and some algebraic properties of the cohomology theory.

Similarly, we can describe the characteristic classes of real vector bundles using geometric properties of the structure group $\mathbf{O}(n)$. However, as we shall use real bundles very little in the following we shall omit some details and refer the reader to other books (Milnor and Stasheff, Husemoller) for further information.

Unlike $\mathbf{U}(n)$, the nontrivial elements of finite order in $\mathbf{O}(n)$ all have order two. In studying $H^*\left(\mathbf{BO}(n)\right)$, it is convenient to describe two separate rings: $H^*\left(\mathbf{BO}(n); \mathbf{Q}\right)$ and $H^*\left(\mathbf{BO}(n); \mathbf{Z}_2\right)$.

Theorem 19 *1. The ring $H^*\left(\mathbf{BSO}(2n); \mathbf{Q}\right)$ is isomorphic to the ring of polynomials $\mathbf{Q}[p_1, p_2, \ldots, p_{n-1}, \chi]$, where*

$$p_k \in H^{4k}\left(\mathbf{BSO}(2n); \mathbf{Q}\right), \ \chi \in H^{2n}\left(\mathbf{BSO}(2n); \mathbf{Q}\right).$$

2. The ring $H^*\left(\mathbf{BS} = \mathbf{O}(2n+1); \mathbf{Q}\right)$ is isomorphic to the ring of polynomials $\mathbf{Q}[p_1, p_2, \ldots, p_n]$, where

$$p_k \in H^{4k}\left(\mathbf{BSO}(2n+1); \mathbf{Q}\right).$$

3. Generators $\{p_k\}$ can be chosen such that the natural mapping

$$\mathbf{BSO}(2n) \longrightarrow \mathbf{BSO}(2n+1)$$

induces a homomorphism such that p_k goes to p_k if $1 \le k \le n-1$ and p_n goes to χ^2. The natural mapping

$$\mathbf{BSO}(2n+1) \longrightarrow \mathbf{BSO}(2n+2)$$

induces a homomorphism such that p_k goes to p_k, $1 \le k \le n$, and χ goes to zero.

4. For a direct sum of bundles:

$$p_k(\xi \oplus \eta) = \sum_{\alpha+\beta=k} p_\alpha(\xi)p_\beta(\eta),$$

where $p_0 = 1$.

Proof.

The group $\mathbf{SO}(2)$ is isomorphic to the circle $\mathbf{S}^1 = \mathbf{U}(1)$. Therefore, the cohomology ring of $\mathbf{BSO}(2)$ is known:

$$H^*(\mathbf{BSO}(2);\ \mathbf{Q}) = \mathbf{Q}[\chi],\ \chi \in H^2(\mathbf{BSO}(2);\ \mathbf{Q}).$$

Consider the bundle

$$\mathbf{BSO}(2k-2) \longrightarrow \mathbf{BSO}(2k) \qquad (2.45)$$

with the fiber $\mathbf{SO}(2k)/\mathbf{SO}(2k-2) = \mathbf{V}_{2k,2}$. The Stieffel manifold $\mathbf{V}_{2k,2}$ is the total space of a bundle

$$\mathbf{V}_{2k,2} \longrightarrow \mathbf{S}^{2k-1} \qquad (2.46)$$

with fiber $\mathbf{S}^{2k-2}$. The manifold $\mathbf{V}_{2k,2}$ may be described as the family of pairs of orthogonal unit vectors (e_1, e_2). We think of the first vector e_1 as parametrising the sphere $\mathbf{S}^{2k-1}$ and the second vector e_2 as tangent to the sphere $\mathbf{S}^{2k-1}$ at the point e_1. Since the sphere $\mathbf{S}^{2k-1}$ is odd dimensional, there is a nontrivial vector field, that is, a section of the bundle (2.46) inducing a monomorphism

$$H^*(\mathbf{S}^{2k-1}) \longrightarrow H^*(\mathbf{V}_{2k,2}).$$

In the spectral sequence for the bundle (2.46) the only one possible nontrivial differential d_{2k} vanishes. Hence

$$E_2^{*,*} = e_\infty^{*,*} = H^*(\mathbf{V}_{2k,2}) = \Lambda(a_{2k-1}, a_{2k-2}).$$

Consider the spectral sequence for the bundle (2.45) (see 2.5).

We take the inductive assumption that

$$H^*(\mathbf{BSO}(2k-2);\ \mathbf{Q}) = \mathbf{Q}[p_1, \ldots, p_{k-2}, \chi], \qquad (2.47)$$

where $\chi \in H^{2k-2}(\mathbf{BSO}(2k-2);\ \mathbf{Q})$. In particular, this means that the odd dimensional cohomology of $\mathbf{BSO}(2k-2)$ is trivial. If

$$d_{2k-1}(a_{2k-2}) = v \neq 0,$$

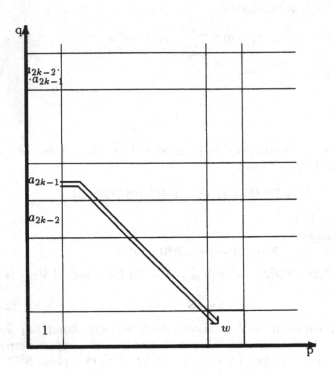

Figure 2.5

then
$$d_{2k-1}(a_{2k-2}v) = v^2 = 0.$$

Hence
$$E_\infty^{2k-1,2k-2} \neq 0.$$

The latter contradicts the assumption that
$$H^{4k-3}(\mathbf{BSO}(2k-2); \mathbf{Q}) = 0.$$

Therefore, one has that
$$d_{2k-1}(a_{2k-2}) = 0.$$

and hence
$$d_{2k-1} = 0.$$

The dimension of the element a_{2k-1} is odd and hence

$$d_{2k}(a_{2k-1}) = w \neq 0.$$

Proceeding as in the proof of Theorem 18, one can check that the ring $H^*(\mathbf{BSO}(2k); \mathbf{Q})$ has only even dimensional elements and the differential d_{2k} maps the subgroup $a_{2k-1}E_{2k}^{*,*}$ monomorphically onto the subgroup $wE_{2k}^{*,*}$. Therefore,

$$E_\infty^{*,*} = E_\infty^{*,0} \oplus a_{2k-2}E_\infty^{*,0}$$

which is associated to the polynomial ring (2.47). This means that

$$\begin{aligned}
p_1,\ldots,p_{k-2} &\in E_\infty^{*,0}, \\
[\chi] &= a_{2k-2}, \\
[\chi^2] &\in E_\infty^{*,0}, \\
E_\infty^{*,0} &= \mathbf{Q}[p_1,\ldots,p_{k-2},\chi^2], \\
E_2^{*,0} &= \mathbf{Q}[p_1,\ldots,p_{k-2},\chi^2,w].
\end{aligned}$$

Put $p_{k-1} = \chi^2$. Then

$$H^*(\mathbf{BSO}(2k); \mathbf{Q}) = \mathbf{Q}[p_1,\ldots,p_{k-1},w].$$

Consider the bundle

$$\mathbf{BSO}(2k-2)\longrightarrow\mathbf{BSO}(2k-1) \tag{2.48}$$

with the fiber $\mathbf{S}^{2k-2}$. In the spectral sequence for the bundle (2.48), the member $E_2^{*,*}$ has only two nontrivial rows: $E_2^{*,0}$ and $E_2^{*,2k-2} = a_{2k-2}E_2^{*,0}$. If

$$d_{2k-1}(a_{2k-2}) = v \neq 0,$$

then

$$d_{2k-1}(a_{2k-2}v) = v^2 = 0.$$

Hence

$$E_\infty^{2k-1,2k-2} \neq 0.$$

and the latter contradicts the previous calculation since the odd dimensional cohomology of the space $\mathbf{BSO}(2k-2)$ is trivial. This means that

$$\begin{aligned}
d_{2k-1}(a_{2k-2}) &= 0, \\
d_{2k-1} &= 0, \\
E_2^{*,*} &= E_\infty^{*,*}.
\end{aligned}$$

Hence the odd dimensional cohomology of the space $\mathbf{BSO}(2k-1)$ is trivial.

Consider the bundle

$$\mathbf{BSO}(2k-1) \longrightarrow \mathbf{BSO}(2k) \qquad\qquad (2.49)$$

with fiber $\mathbf{S}^{2k-1}$ The member $E_2^{*,*}$ of the spectral sequence of the bundle (2.49) has the form

$$E_2^{*,*} = H^*\left(\mathbf{BSO}(2k)\right) \otimes \Lambda(a_{2k-1}).$$

The odd dimensional cohomology of the total space $\mathbf{BSO}(2k-1)$ is trivial and so

$$d_{2k}(a_{2k-1}) = v \neq 0.$$

Using the calculation for the bundle (2.45), we see that

$$v = d_{2k}(a_{2k}) = w.$$

Hence

$$d_{2k}(a_{2k-1}x) = wx.$$

Thus

$$E_{2k+1}^{*,*} = E_\infty^{*,*} = \mathbf{Q}[q_1\ldots,p_{k-1}] = H^*\left(\mathbf{BSO}(2k-1);\ \mathbf{Q}\right).$$

To prove the last relation of Theorem 19 we study the bundle

$$j_n : \mathbf{BSO}(2) \times \ldots \times \mathbf{BSO}(2) \longrightarrow \mathbf{BSO}(2n).$$

We need to prove that the image of j_n^* coincides with the ring of symmetric polynomials of variables $\{t_1^2, p_2^2, \ldots, p_n^2\}$ where

$$t_k \in H^*\left(\mathbf{BSO}(2) \times \ldots \times \mathbf{BSO}(2);\ \mathbf{Q}\right)$$

are generators. We leave this proof for reader.   ■

The characteristic classes

$$p_k \in H^{4k}\left(\mathbf{BSO}(n);\ \mathbf{Q}\right)$$

constructed in Theorem 19 are called *the rational Pontryagin classes* . The class

$$\chi \in H^{2n}\left(\mathbf{BSO}(2n);\ \mathbf{Q}\right)$$

is called *the Euler class*

Theorem 20 *The ring $H^*\left(\mathbf{BO}(n);\ \mathbf{Z}_2\right)$ is isomorphic to the polynomial ring* $\mathbf{Z}[w_1,\ldots,w_n]$,

$$H^*\left(\mathbf{BO}(n);\ \mathbf{Z}_2\right) = \mathbf{Z}[w_1,\ldots,w_n],\ w_k \in H^k\left(\mathbf{BO}(n);\ \mathbf{Z}_2\right).$$

The generators w_k can be chosen such that

1.

$$w_k(\xi) = 0\ \textit{for}\ k > \dim\xi;$$

2.

$$w_k(\xi \oplus 1) = w_k(\xi);$$

3.

$$w_k(\xi \oplus \eta) = \sum_{\alpha+\beta=k} w_\alpha(\xi) w_\beta(\eta).$$

The proof of this theorem is left to the reader. The class w_k is called *Stieffel–Whitney characteristic class* .

Exercise

Prove that the structure group $\mathbf{O}(n)$ of the vector bundle ξ can bes reduced to the subgroup $\mathbf{SO}(n)$ if and only if $w_1(\xi) = 0$. For this consider the bundle

$$\mathbf{BSO}(n) \longrightarrow \mathbf{BO}(n),$$

a two–sheeted covering. If $w_1(\xi) = 0$ then the vector bundle ξ is said to be an *oriented bundle* .

Example

Consider the space $\mathbf{BO}(1) = \mathbf{RP}^\infty$ and the universal one–dimensional Hopf vector bundle over it. It is clear that

$$w_1(\xi) = w_1 \in H^1\left(\mathbf{RP}^\infty;\ \mathbf{Z}_2\right)$$

is a generator and

$$w_1(\xi \oplus \xi) = w_1(\xi) + w_1(\xi) = 2w_1 = 0.$$

Hence the bundle $\xi \oplus \xi$ is orientable. The bundle $\xi \oplus \xi$ is nontrivial since

$$w_2(\xi \oplus \xi) = w_1(\xi)w_1(\xi) = w_1^2 \neq 0.$$

2.5   GEOMETRIC INTERPRETATION OF SOME CHARACTERISTIC CLASSES

2.5.1   The Euler class

Let X be a smooth, closed, oriented manifold of dimension $2n$. We put the question: Is there a nonvanishing vector field on the manifold X? Let TX be the tangent bundle and SX the bundle consisting of all unit tangent vectors. Then the existence of a nonvanishing vector field is equivalent to the existence of a section of the bundle SX. The bundle

$$SX \longrightarrow X \tag{2.50}$$

with the fiber $\mathbf{S}^{2n-1}$ is classified by a mapping

$$X \longrightarrow \mathbf{BSO}(2n) \tag{2.51}$$

and the universal bundle

$$\mathbf{SESO}(2n) \longrightarrow \mathbf{BSO}(2n)$$

with the fiber $\mathbf{S}^{2n-1}$.

The classifying space $\mathbf{BSO}(2n)$ can be taken as the space of $2n$–dimensional oriented linear subspaces of an infinite dimensional vector space. The points of the total space $\mathbf{SESO}(2n)$ can be represented by pairs (L, l) where L is $2n$–dimensional subspace and $l \in L$ is a unit vector.

Let $l^\perp \subset L$ be the orthonormal supplement to the vector l. Then the correspondence

$$(L, l) \longrightarrow l^\perp$$

gives a locally trivial bundle

$$\mathbf{SESO}(2n) \longrightarrow \mathbf{BSO}(2n-1). \tag{2.52}$$

The fiber of this bundle is an infinite dimensional sphere. Hence the mapping (2.52) is a homotopy equivalence and the bundle (2.50) is classified by the bundle

$$\mathbf{BSO}(2n-1) \longrightarrow \mathbf{BSO}(2n), \tag{2.53}$$

with the fiber $\mathbf{S}^{2n-1}$.

Let $E_s^{p,q}$ be the spectral sequence for the bundle (2.53) and $E_s^{p,q}(X)$ be the spectral sequence for the bundle (2.50). Let

$$f : E_s^{p,q} \longrightarrow E_s^{p,q}(X)$$

be the homomorphism induced by the mapping (2.51). Then the element

$$f^*(\chi) \in H^{2n}(X) = E_2^{2n,0}(X)$$

is the Euler class of the manifold X. In the section 2.4 (Theorem 19), it was shown that projection

$$p : \mathbf{BO}(2n-1) \longrightarrow \mathbf{BO}(2n)$$

sends the Euler class to zero. Hence the same holds for the projection

$$p' : SX \longrightarrow X$$

where:

$$p'^*(f^*(\chi)) = 0.$$

If

$$\alpha : X \longrightarrow SX$$

is a section, then

$$0 = \alpha^* p'^*(f^*(\xi)) = f^*(\chi). \tag{2.54}$$

Thus the existence of nonvanishing vector field on the manifold X implies that the Euler class of the manifold X is trivial.

2.5.2   Obstruction to constructing a section

In section 2.4 the Euler class was defined as a cohomology class with rational coefficients. But it is possible to define it for integer cohomology. As a matter of fact, the triviality of the integer Euler class is sufficient for the existence of nonvanishing vector field on an oriented $2n$–dimensional manifold. On the other hand, the triviality of the rational Euler class implies the triviality of the integer Euler class since $2n$–dimensional cohomology group has no torsion.

Generally speaking, there are homology type obstructions to the existence of a section of a locally trivial bundle. Some of them may be described in terms of characteristic classes.

Let
$$p : E \longrightarrow X$$
be a locally trivial bundle with fiber F and let X be a cellular space. Assume that there exists a section
$$s : [X]^{n-1} \longrightarrow E$$
over $(n-1)$–dimensional skeleton of X. We can ask whether it is possible to extend the section s over the n–dimensional skeleton $[X]^n$. It is sufficient to extend the section s over each n–dimensional cell $\sigma_i \subset X$ separately. Over a cell σ_i the bundle is trivial, that is,

$$p^{-1}(\sigma_i) \equiv \sigma_i \times F,$$

the restriction of the section s to the boundary $\partial \sigma_i$ defines a mapping to the fiber F:
$$s : \partial \sigma_i : \longrightarrow F,$$
and which defines an element of the homotopy group of the fiber:
$$[s]_i \in \pi_{n-1}(F).$$

This element will be well defined, that is, it does not depend on the choice of the fiber and on the choice of fixed point of F if both the base and fiber are simply connected.

Thus necessary and sufficient conditions for the existence of an extension of the section s to the skeleton $[X]^n$ are given by the following relations:
$$[s]_i = 0 \in \pi_{n-1}(F).$$

In other words, the cochain
$$[s] \in C^n(X, \pi_{n-1}(F))$$

defined by the section s may be considered as a obstruction to the extension of the section s to the n–dimensional skeleton $[X]^n$.

The cochain $[s]$ is not arbitrary. In fact, it is cocycle:
$$\partial[s] = 0.$$

To prove this, consider an $(n+1)$–dimensional cell Σ. Let
$$f : \partial \Sigma \longrightarrow [X]^n$$

be the mapping which defines the gluing of the cell to the skeleton $[X]^n$. Consider the composition

$$\tilde{f} : \partial\Sigma \longrightarrow [X]^n \longrightarrow [X]^n/[X]^{n-1} = \vee \sigma_i/\partial\sigma_i.$$

Using a homotopy of the mapping f we can assume that $\tilde{f}$ is smooth everywhere except at the fixed point x_0 of $\vee\sigma_i/\partial\sigma_i$ and the points $x_i \in \sigma_i/\partial\sigma_i$ are nonsingular. This means that

$$\tilde{f}^{-1}(x_0) = \cup_{i,j} D_{ij}$$

where the $D_{i,j} \subset \partial\Sigma$ are n-dimensional disks which are pairwise disjoint and for which each

$$\tilde{f}_{|D_{i,j}} : D_{i,j} \longrightarrow \sigma_i/\partial\sigma_i \qquad (2.55)$$

is a diffeomorphism on the interior of $D_{i,j}$. Let $t_{i,j} = \pm 1$ be the degree of the mapping (2.55). Then for the cell chain generated by single cell Σ we have the boundary homomorphism

$$\delta\Sigma = \sum_i \left(\sum_j t_{i,j} \right) \sigma_i.$$

with corresponding adjoint boundary homomorphism on cochains

$$(\partial x)(\Sigma) = \sum_{i,j} t_{i,j} x(\sigma_i). \qquad (2.56)$$

Now the section s is defined on the $\partial\Sigma \backslash \cup_{i,j} \int D_{i,j}$. The restriction of the section s to $\partial D_{i,j}$ gives an element $[s]_{i,j}$ of $\pi_{n-1}(F)$ for which

$$[s]_{i,j} = t_{i,j}[s]_i \in \pi_{n-1}(F).$$

Thus from (2.56),

$$(\partial[s])(\Sigma) = \sum_{i,j} t_{i,j}[s](\sigma_i) = \sum_{i,j} t_{i,j}[s]_i = \sum_{i,j}[s]_{i,j}.$$

Lemma 5 *Let $D_k \subset \partial\Sigma$ be a family of disjoint disks and let*

$$g : \partial\Sigma \backslash \cup_k \int D_k \longrightarrow F$$

be a continuous mapping which on each ∂D_k gives an element $[g]_k \in \pi_{n-1}(F)$. Then

$$\sum_k [g]_k = 0 \in \pi_k(F).$$

Lemma 5 finishes proof that $[s]$ is cocycle.

The obstruction theory says that the cocycle $[s] \in C^n(X, \pi_{n-1}(F))$ gives a cohomology class $[s] \in H^n(X; \pi_{n-1}(F))$ called the *obstruction to extending the section* $[s]$. Further, if this obstruction is trivial then there is another section $[s']$ on the skeleton $[X]^n$ which coincides with $[s]$ on the skeleton $[X]^{n-2}$. If n is smallest number for which the group $H^n(X; \pi_{n-1}(F))$ is nontrivial then the corresponding obstruction $[s]$ is called *the first obstruction to the construction a section*.

Theorem 21 *Let* $p : E \longrightarrow X$ *be locally trivial bundle with the fiber* F *and let* X *be a simply connected cellular space. Let*

$$\pi_0(F) = \dots \pi_{n-2}(F) = 0$$
$$\pi_{n-1}(F) = \mathbf{Z}.$$

Let $a \in H^{n-1}(F; \mathbf{Z})$ *be a generator and let* $E_s^{p,q}$ *be the spectral sequence of the bundle* p. *Then the first nontrivial differential acting on the element* a *is* d_n, $E_n^{n,0} = E_2^{n,0} = H^n(X; \mathbf{Z})$, *and the element* $w = d_n(a) \in H^n(X; \mathbf{Z})$ *is the first obstruction to constructing a section of the bundle* p.

Theorem 22 *Let* ξ *be a complex* n-*dimensional vector bundle over a base* X. *Then the Chern class* $c_k(\xi) \in H^{2k}(X; \mathbf{Z})$ *coincides with the first obstruction to constructing of* $n - k + 1$ *linear independent sections of the bundle* ξ.

Proof.

Constructing $n - k + 1$ linearly independent sections of ξ is the same as representing the bundle ξ as a direct sum

$$\xi = \eta \oplus \overline{n - k + 1}. \tag{2.57}$$

Consider the bundle
$$p : \mathbf{BU}(k - 1) \longrightarrow \mathbf{BU}(n) \tag{2.58}$$
with the fiber $\mathbf{U}(n)/\mathbf{U}(k - 1)$. Let

$$f : X \longrightarrow \mathbf{BU}(n)$$

be a continuous mapping inducing the bundle ξ, that is,

$$\xi = f^*(\xi_n).$$

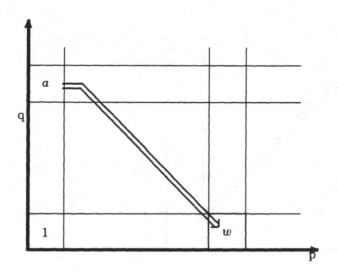

Figure 2.6

Representing the bundle as in (2.57) means constructing a mapping

$$g : X \longrightarrow \mathbf{BU}(k-1)$$

such that $pg \sim f$. In other words, the first obstruction to constructing $n-k+1$ linearly independent sections of the bundle ξ coincides with the inverse image of the first obstruction to constructing a section of the bundle (2.58). Let us use Theorem 21.

Consider the spectral sequence for the bundle (2.58) or the figure (2.7). Since $p^*(c_k) = 0$, the element $c_k \in E_2^{2k,0}$ is the image of the differential d_{2k},

$$d_{2k}(u_{2k-1}) = c_k.$$

By Theorem 21, we have

$$f^*(c_k) = c_k(\xi)$$

giving the first obstruction to constructing $n-k+1$ linearly independent sections of the bundle x. ∎

There are similar interpretations for the Stieffel–Whitney classes and the Pontryagin classes.

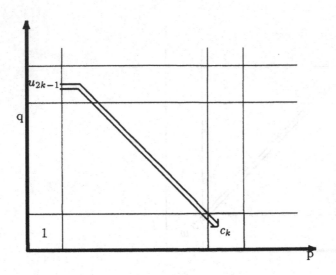

Figure 2.7

2.6 K–THEORY AND THE CHERN CHARACTER

2.6.1 Definitions and simple properties

The family of (isomorphism classes of) vector bundles over a fixed base is not a group with respect to the operation of direct sum since subtraction does not in general exist for any two bundles. However this family may be extended to form a group by introducing a formal subtraction. If the base is not connected we will consider a vector bundle as a union vector bundles of arbitrary dimensions, one on each connected component.

Definition 6 *Let $K(X)$ denotes the abelian group where the generators are (isomorphism classes of) vector bundles over the base X subject to the following relations:*

$$[\xi] + [\eta] - [\xi \oplus \eta] = 0 \qquad (2.59)$$

for vector bundles ξ and η, and where $[\xi]$ denotes the element of the group $K(X)$ defined by the vector bundle ξ.

The group defined in Definition 6 is called *the Grothendieck group* of the category of all vector bundles over the base X.

To avoid confusion, the group generated by all real vector bundles will be denoted by $K_O(X)$, the group generated by all complex vector bundles will be denoted by $K_U(X)$ and the group generated by all quaternionic vector bundles will be denoted by $K_{Sp}(X)$.

The relations (2.59) show that in the group $K(X)$ the direct sum corresponds to the group operation.

Proposition 7 *Each element $x \in K(X)$ can be represented as a difference of two vector bundles:*

$$x = [\zeta_1] - [\zeta_2].$$

Proof.

In the general case, the element x is a linear combination

$$x = \sum_{i=1}^{p} n_i[\xi_i]$$

with the integer coefficients n_i. Let us split this sum into two parts

$$x = \sum_{i=1}^{p_1} n_i[\xi_i] - \sum_{j=1}^{p_2} m_j[\eta_j]$$

with

$$n_i > 0,\ m_j > 0.$$

Using the relations (2.59) we have

$$x = [\oplus_{i=1}^{p_1} \underbrace{(\xi_i \oplus \ldots \oplus \xi_i)}_{n_i \text{ times}}] - [\oplus_{j=1}^{p_2} \underbrace{(\eta_j \oplus \ldots \oplus \eta_j)}_{m_j \text{ times}}] =$$

$$= [\zeta_1] - [\zeta_2].$$

∎

Note: In Proposition 7 the bundle ζ_2 may be taken to be trivial. In fact, using Theorem 8 from the chapter 1 there is a vector bundle ζ_3 such that

$$\zeta_2 \oplus \zeta_3 = \bar{N}$$

is a trivial bundle. Then

$$x = [\zeta_1] - [\zeta_2] = ([\zeta_1] - [\zeta_2]) + ([\zeta_3] - [\zeta_{=}3]) =$$
$$= \ ([\zeta_1] + [\zeta_3]) - ([\zeta_2] + [\zeta_3]) = [\zeta_2 \oplus \zeta_3] - \bar{N}.$$

■

Proposition 8 *Two vector bundles ξ_1 and ξ_2 define the same element in the group $K(X)$ if and only if there is a trivial vector bundle $\bar{N}$ such that*

$$\xi_1 \oplus \bar{N} = \xi_2 \oplus \bar{N}. \tag{2.60}$$

Condition 2.60 is called *stable isomorphism of vector bundles* . **Proof.**

If the condition (2.60) holds then by (2.59)

$$[\xi_1] + [\bar{N}] = [\xi_2] + [\bar{N}],$$

that is, $[\xi_1] = [\xi_2]$. Conversely, assume that $[\xi_1] = [\xi_2]$ or $[\xi_1] - [\xi_2] = 0$. Then by definition, this relation can be represented as a linear combination of determining relations:

$$[\xi_1] - [\xi_2] = \sum_j \lambda_j (\eta_j + \zeta_j - \eta_j \oplus \zeta_j). \tag{2.61}$$

Without loss of generality we can consider that $\lambda_j \pm 1$. We transform the identity (2.61) in the following way: all summands with negative coefficients carry over to the other side of the equality. Then we have

$$\xi_1 + \sum_j (\eta_j \oplus \zeta_j) + \sum_k (\eta_k + \zeta_k) =$$
$$= \ \xi_2 + \sum_j (\eta_j + \zeta_j) + \sum_k (\eta_k \oplus \zeta_k) \tag{2.62}$$

where j runs through all indices for which $\lambda_j = 1$ and k runs through all indices for which $\lambda_k = -1$. Both parts of the equality (2.62) are formal sums of vector bundles. Hence each summand on the left hand side is isomorphic to a summand on the right hand side and vice versa. The direct sum of all summands in the left hand side is equal to the direct sum of all summand in the right hand side. Thus

$$\xi_1 \oplus \sum_j (\eta_j \oplus \zeta_j) \oplus \sum_k (\eta_k \oplus \zeta_k) =$$

$$= \xi_2 \oplus \sum_j (\eta_j \oplus \zeta_j) \oplus \sum_k (\eta_k \oplus \zeta_k).$$

Put

$$\theta = \sum_j (\eta_j \oplus \zeta_j) \oplus \sum_k (\eta_k \oplus \zeta_k).$$

Then we have

$$\xi_1 \oplus \theta = \xi_2 \oplus \theta.$$

Finally, let χ be a vector bundle such that $\bar{N} = \chi \oplus \theta$ then

$$\xi_1 \oplus \bar{N} = \xi_2 \oplus \bar{N}.$$

∎

Proposition 9 *The operation of tensor product of vector bundles induces a ring structure in the additive group $K(X)$.*

We shall leave the proof to the reader.

Examples

1. We describe the ring $K(x_0)$ where x_0 is a one point space. Each vector bundle over a point is trivial and characterized by the dimension of the fiber. Therefore, the family of all vector bundles over a point, with two operations — direct sum and tensor product, is isomorphic to the semiring of positive integers. Hence the ring $K(x_0)$ is isomorphic to the ring $\mathbf{Z}$ of integers. In addition, the difference $[\xi] - [\eta]$ corresponds the number $\dim \xi - \dim \eta \in \mathbf{Z}$.

2. If $X = \{x_1, x_2\}$ is a two point space then any vector bundle is defined by two integers — the dimension of the fiber over the point x_1 and the dimension of the fiber over the point x_2. Hence

$$K(x_0) = \mathbf{Z} \oplus \mathbf{Z}.$$

2.6.2 K–theory as a generalized cohomology theory

The group $K(X)$ is a ring. It is clear that a continuous mapping $f : X \longrightarrow Y$ induces a ring homomorphism

$$f^* : K(Y) \longrightarrow K(X)$$

which to each vector bundle ξ associates the inverse image $f^*(\xi)$. If

$$f, g : X \longrightarrow Y$$

are two homotopic mappings then

$$f^* = g^* : K(Y) \longrightarrow K(X).$$

In fact, there is a continuous map

$$F : X \times I \longrightarrow Y$$

such that

$$F(x, 0) = f(x); \; F(x, 1) = g(x).$$

Then by Theorem 3 from the chapter 1, restrictions of the bundle $F^*(\xi)$ to the subspaces $X \times \{0\}$ and $X \times \{1\}$ are isomorphic. Hence

$$f^*(\xi) = g^*(\xi).$$

Therefore, if two spaces X and Y are homotopy equivalent then the rings $K(X)$ and $K(Y)$ are isomorphic.

This means that the correspondence

$$X \longrightarrow K(X)$$

is a homotopy functor from the category of homotopy types to the category of rings. One can extend this functor to an analogue of a generalized cohomology theory. Let (X, x_0) be a cellular space with a base point. The natural inclusion $x_0 \longrightarrow X$ induces a homomorphism of rings:

$$K(X) \longrightarrow K(x_0) = \mathbf{Z}. \tag{2.63}$$

This homomorphism is defined by a formula for a difference $[\xi] - [\eta]$:

$$[\xi] - [\eta] \longrightarrow \dim \xi - \dim \eta \in \mathbf{Z}$$

Note that if X has many connected components then we take the dimension of the bundle x over the component containing the base point x_0. Therefore, the homomorphism (2.63) will also be denoted by $\dim : K(X) \longrightarrow K(x_0)$.

Let $K^0(X, x_0)$ denote the kernel of the homomorphism (2.63):

$$K^0(X, x_0) = \mathbf{Ker}\ (K(X) \longrightarrow K(x_0))\,.$$

Elements of the subring $K^0(X, x_0)$ are represented by differences $[\xi] - [\eta]$ for which $\dim \xi = \dim h$. The elements of the ring $K(X)$ are called *virtual* bundles and elements of the ring $K^0(X, x_0)$ are *virtual* bundles of trivial dimension over the point x_0.

Now consider a pair (X, Y) of the cellular spaces, $Y \subset X$. Denote by $K^0(X, Y)$ the ring

$$K^0(X, Y) = K^0(X/Y, [Y]).$$

For any negative integer $-n$, let

$$K^{-n}(X, Y) = K^0(S^n X, S^n Y)$$

where $S^n(X)$ denotes the n-times suspension of the space X:

$$S^n X = (S^n \times X)/(S^n \vee X).$$

Theorem 23 *The pair (X, Y) induces an exact sequence*

$$K^0(Y, x_0) \longleftarrow K^0(X, x_0) \longleftarrow K^0(X, Y) \longleftarrow$$
$$\longleftarrow K^{-1}(Y, x_0) \longleftarrow K^{-1}(X, x_0) \longleftarrow K^{-1}(X, Y) \longleftarrow$$
$$\longleftarrow K^{-2}(Y, x_0) \longleftarrow K^{-2}(X, x_0) \longleftarrow K^{-2}(X, Y) \longleftarrow \ldots$$

$$\cdots \ \longleftarrow K^{-n}(Y, x_0) \longleftarrow K^{-n}(X, x_0) \longleftarrow K^{-n}(X, Y) \longleftarrow \cdots$$

$$\tag{2.64}$$

Proof.

The sequence (2.64) is based on the so called Puppe sequence of spaces and may be is constructed for any pair (X, Y):

$$Y \longrightarrow X \longrightarrow X/Y \longrightarrow$$
$$\longrightarrow S^1 Y \longrightarrow S^1 X \longrightarrow S^1(X/Y) \longrightarrow$$

$$\longrightarrow S^2 Y \longrightarrow S^2 X \longrightarrow S^2(X/Y) \longrightarrow \dots$$

$$\dots \longrightarrow S^n Y \longrightarrow S^n X \longrightarrow S^n(X/Y) \longrightarrow \dots$$

In the Puppe sequence, all continuous mapping may be taken as inclusions by substituting the gluing of a cone for each quotient space or factorization. Then the next space in the sequence can be considered as a quotient space. Hence to prove exactness it is sufficient to check exactness of the sequence

$$K^0(Y, x_0) \xleftarrow{i^*} K^0(X, x_0) \xleftarrow{j^*} K^0(X, Y)$$

at the middle term. Let ξ be a vector bundle over X for which the restriction to the subspace Y is trivial. This means that the bundle ξ is trivial in some neighborhood U of the Y. Hence the neighborhood U can serve as a chart of an atlas for ξ. In addition, other charts can be chosen so that they do not intersect Y. Hence all the transition functions are defined away from the subspace Y. This means that the same transition functions define transition functions on the quotient space X/Y, that is, they define a vector bundle η over X/Y such that

$$j^*([\eta]) = [\xi].$$

∎

Remark

We have not quite obtained a generalized cohomology theory but only the part graded by nonnegative integers.

2.6.3 The Chern character

In the section 2.4 we defined Chern classes for any complex vector bundle. By Theorem 18, an arbitrary characteristic class is a polynomial of the Chern classes $c_1(\xi), \dots, c_n(\xi)$ with integer coefficients. If we consider characteristic classes in cohomology with rational coefficients then they may all be expressed as polynomials in Chern classes with rational coefficients.

Consider now the function

$$\varphi^n(t_1, \dots, t_n) = e^{e_1} + e^{t_2} + \dots + e^{t_n}$$

and its the Taylor series

$$\varphi^n(t_1, \ldots, t_n) = \sum_{k=0}^{\infty} \frac{t_1^k + \ldots + t_n^k}{k!}. \tag{2.65}$$

Each term of the series (2.65) is a symmetric polynomial in the variables $t_1, \ldots, t_n$ and hence may be expressed as a polynomial in the elementary symmetric polynomials

$$\frac{t_1^k + \ldots + t_n^k}{k!} = \varphi_k^n(\sigma_1, \ldots, \sigma_n),$$
$$\sigma_1 = \sigma_1(t_1, \ldots, t_n) = t_1 + t_2 + \ldots + t_n,$$
$$\cdots\cdots\cdots\cdots\cdots \quad \cdots \quad \cdots\cdots\cdots\cdots$$
$$\sigma_n = \sigma_n(t_1, \ldots, t_n) = t_1 t_2 \cdots t_n,$$

In other words,

$$\varphi^n(t_1, \ldots, t_n) = \sum_{k=0}^{\infty} \varphi_k^n(\sigma_1, \ldots, \sigma_n).$$

Definition 7 *Let ξ be a n–dimensional complex vector bundle over a space X. The characteristic class*

$$ch\, \xi = \sum_{k=0}^{\infty} \sum_{k=0}^{\infty} \varphi_k^n(c_1(\xi), \ldots, c_n(\xi)) \tag{2.66}$$

is called the Chern character of the bundle ξ.

In the formula (2.66), each summand is an element of the cohomology group $H^{2k}(X, \mathbf{Q})$. If the space X is a finite cellular space then the Chern character $ch\, \xi$ is a nonhomogeneous element of the group

$$H^{2*}(X; \mathbf{Q}) = \oplus H^{2k}(X; \mathbf{Q})$$

since from some number k_0 on all the cohomology groups of the space X are trivial. In general, we consider that the formula (2.66) defines an element of the group

$$H^{**}(X; \mathbf{Q}) = \prod_{k=0}^{\infty} H^k(X; \mathbf{Q}).$$

Theorem 24 *The Chern character satisfies the following conditions:*

1. $ch\ (\xi \oplus \eta) = ch\ (\xi) + ch\ (\eta)$;

2. $ch\ _0(\xi) = \dim \xi$;

3. $ch\ (\xi \otimes \eta) = ch\ (\xi)ch\ (\eta)$.

Proof.

As in the section 2.4, we can represent the Chern classes by elementary symmetric polynomial in certain variables, that is,

$$
\begin{aligned}
c_k(\xi) &= \sigma_k(t_1,\ldots,t_n), \\
c_k(\eta) &= \sigma_k(t_{n+1},\ldots,t_{n+m}), \\
c_k(\xi \oplus \eta =) &= \sigma_k(t_1,\ldots,t_n,t_{n+1},\ldots,t_{n+m}).
\end{aligned}
\tag{2.67}
$$

Using (2.66),

$$
\begin{aligned}
\text{ch}\ _k(\xi) &= \varphi_k^n(c_1(\xi),\ldots,c_n(\xi)) = \frac{1}{k!}\left(t_1^k + \cdots + t_n^k\right), \\
\text{ch}\ _k(\eta) &= \varphi_k^m(c_1(\eta),\ldots,c_n(\eta)) = \frac{1}{k!}\left(t_{n+1}^k + \cdots + t_{n+m}^k\right), \\
\text{ch}\ _k(\xi \oplus \eta) &= \varphi_k^{n+m}(c_1(\xi \oplus \eta =),\ldots,c_{n+m}(\xi \oplus \eta)) = \\
&= \frac{1}{k!}\left(t_1^k + \cdots + t_n^k + t_{n+1}^k + \cdots + t_{n+m}^k\right) = \\
&= \text{ch}\ _k(\xi) + \text{ch}\ _k(\eta).
\end{aligned}
\tag{2.68}
$$

and from (2.68) property 1 follows. Property 2 follows from

$$
\text{ch}\ _0(\xi) = \varphi_0^n(c_1(\xi),\ldots,c_n(\xi)) = n = \dim \xi.
$$

Strictly speaking this proof is not quite correct because we did not make sense of the formulas (2.67). We can make sense of them in the following way. For all three identities in Theorem 24 it is sufficient to check them over the special spaces

$$
X = \mathbf{BU}(n) \times \mathbf{BU}(m),
$$

where $\xi = \xi_n$ is the universal vector bundle over the factor $\mathbf{BU}(n)$ and $\eta = \xi_m$ is the universal vector bundle over the factor $\mathbf{BU}(m)$. Let

$$
f : \mathbf{BU}(n) \times \mathbf{BU}(m) \longrightarrow \mathbf{BU}(n + m)
$$

be a mapping such that

$$
f^*(\xi_{n+m}) = \xi_n \oplus \xi_m.
$$

Then we must prove that

$$f^*(\mathrm{ch}\ \xi_{n+m}) = \mathrm{ch}\ \xi_n + \mathrm{ch}\ \xi_m. \tag{2.69}$$

Consider a commutative diagram

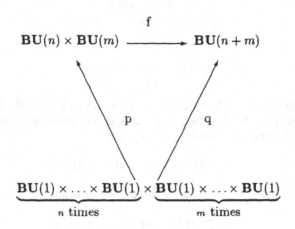

where the mappings p, q correspond to the inclusions of the group of diagonal matrices into the unitary groups. In the notation of section 2.4,

$$p = j_n \times j_m, \quad q = j_{n+m}.$$

By the lemma 4 both homomorphisms p^* and q^* are monomorphisms on cohomology. Hence the identity (2.69) follows from the following identity

$$p^* f^*(\mathrm{ch}\ \xi_{n+m}) = p^* \mathrm{ch}\ \xi_n + p^* \mathrm{ch}\ \xi_m,$$

that is, from

$$q^*(\mathrm{ch}\ \xi_{n+m}) = p^* \mathrm{ch}\ \xi_n + p^* \mathrm{ch}\ \xi_m$$

or from

$$\mathrm{ch}\ q^*(\xi_{n+m}) = \mathrm{ch}\ p^* \xi_n + \mathrm{ch}\ p^* \xi_m.$$

Let the η_i be one dimensional vector bundles, $1 \leq i \leq n+m$, which are universal vector bundles over the various factors with corresponding number i. Then

$$
\begin{aligned}
q^* \xi_{n+m} &= \eta_1 \oplus \ldots \oplus \eta_{n+m}, \\
p^* \xi_n &= \eta_1 \oplus \ldots \oplus \eta_n, \\
p^* \xi_m &= \eta_{n+1} \oplus \ldots \oplus \eta_{n+m}.
\end{aligned}
$$

Therefore, we can put $t_i = c_1(\eta_i)$ and then the formulas (2.67) make sense.

We now pass to the proof of the property 3. As in the previous case it is sufficient to provide the proof for bundles

$$
\begin{aligned}
\xi &= \eta_1 \oplus \ldots \oplus \eta_n \\
\eta &= \eta_{n+1} \oplus \ldots \oplus \eta_{n+m}.
\end{aligned}
$$

Then

$$
\xi \otimes \eta = \bigoplus_{i,j} (\eta_i \otimes \eta_{n+j}).
$$

So we need to calculate the first Chern class $c_1(\xi' \otimes \xi'')$ for the product of two one dimensional vector bundles which might be considered universal bundles over factors the $\mathbf{BU}(1)$. Consider a mapping

$$
f : \mathbf{BU}(1) \times \mathbf{BU}(1) \longrightarrow \mathbf{BU}(1)
$$

such that $f^*(\xi) = \xi' \otimes \xi''$. Then for some integer α and β

$$
c_1(\xi' \otimes \xi'') = \alpha c_1(\xi') + \beta c_1(\xi'').
$$

When $\xi' = 1$ we have $\xi' \otimes \xi'' = \xi''$ and hence

$$
c_1(\xi'') = \alpha c_1(1) + \beta c_1(\xi'') = \beta c_1(\xi''),
$$

that is, $\beta = 1$. Similarly, $\alpha = 1$. Finally,

$$
\begin{aligned}
\mathrm{ch}\,(\xi \otimes \eta) &= \sum_{i,j} e^{c_1(\eta_i \otimes \eta_{n+j})} = \\
&= \sum_{i,j} e^{t_i + t_{n+j}} = \sum_{i,j} e^{t_i} e^{t_{n+j}} = \\
&= \left(\sum_i e^{t_i} \right) \left(\sum_j e^{t_{n+j}} \right) = \mathrm{ch}\,\xi\,\mathrm{ch}\,\eta.
\end{aligned}
$$

Theorem 24 shows that the Chern character can be extended to a homomorphism of rings

$$
\mathrm{ch} : K(X) \longrightarrow H^{2*}(X;\ \mathbf{Q}).
$$

3

GEOMETRIC CONSTRUCTIONS
OF BUNDLES

To extend the study of the properties of the vector bundles we need further geometric ideas and constructions. This chapter is devoted to the most frequently used constructions which lead to deeper properties of vector bundles. They are Bott periodicity – the main instrument of the calculation of K–theory, linear representations and cohomology operations in K–theory and the Atiyah–Singer formula for calculating of the indices of elliptic operators on compact manifolds. The last of these will be considered in the next chapter.

3.1   THE DIFFERENCE CONSTRUCTION

Consider a vector bundle ξ over the base X. Assume that the bundle ξ is trivial over a closed subspace $Y \subset X$. This means that there is a neighborhood $U \supset Y$ such that the restriction of the bundle ξ to U is isomorphic to a Cartesian product:

$$\xi_{|U} \sim U \times \mathbf{R}^n.$$

It is clear that the set U can serve as chart and, without loss of generality, we can consider that the other charts do not intersect with Y. Hence all the transition functions $\varphi_{\alpha\beta}$ are defined at points of the complement $X \backslash Y$. Therefore, the same transition functions can be used for the definition of a bundle η over the new base — the quotient space X/Y. The bundle η can be considered as a bundle obtained from ξ by identification of fibers over Y with respect to the isomorphism $\xi_{|Y} \sim Y \times \mathbf{R}^n$.

This important construction can be generalized in the following way. Consider a pair (X, Y) of cellular spaces and a triple (ξ_1, d, ξ_2) where ξ_1 and ξ_2 are vector bundles over X and d is an isomorphism

$$d : \xi_1|_Y \longrightarrow \xi_2|_Y.$$

In the case when ξ_2 is a trivial bundle we have the situation described above.

Consider the family of all such triples. The triple (ξ_1, d, ξ_2) is said to be trivial if d can be extended to an isomorphism over the whole base X. If two isomorphisms $d_1, d_2 : \xi_1|_Y \longrightarrow \xi_2|_Y$ are homotopic in the class of isomorphisms then the two triples (ξ_1, d_1, ξ_2) and (ξ_1, d_2, ξ_2) are said to be equivalent. If the triple (η_1, h, η_2) is trivial then the triples (ξ_1, d, ξ_2) and $(\xi_1 \oplus \eta_1, d \oplus h, \xi_2 \oplus \eta_2)$ are also said to be equivalent. We denote by $\mathcal{K}^2(X, Y)$ the family of classes of equivalent triples.

Theorem 25 *The family* $\mathcal{K}^2(X, Y)$ *is an Abelian group with respect to the direct sum of triples. The group* $\mathcal{K}^2(X, Y)$ *is isomorphic to* $K^0(X, Y)$ *by the correspondence*

$$K^0(X, Y) \ni ([\xi] - [\bar{N}]) \longrightarrow (p^*\xi, d, \bar{N})$$

where $p : X \longrightarrow X/Y$ *is natural projection and* $d : p^*\xi|_Y \longrightarrow \bar{N}$ *is natural isomorphism generated by trivialization on a chart which contains the point* $[Y] \in X/Y$, *(*$\dim \xi = N$*).*

Proof.

The operation of summation is defined by

$$(\xi_1, d_1, \xi_2) + (\eta_1, d_2, \eta_2) = (\xi_1 \oplus \eta_1, d_1 \oplus d_2, \xi_2 \oplus \eta_2).$$

Hence associativity is clear. It is clear that a trivial triple gives a neutral element. If (ξ_1, d_1, ξ_2) is trivial triple and d is homotopic to d' in the class of isomorphisms then the triple (ξ_1, d'_1, ξ_2) also is trivial. Indeed, assume firstly that both vector bundles ξ_1 and ξ_2 are trivial. Then triviality means that the mapping $d : Y \longrightarrow \mathbf{GL}(n)$ extends to the mapping $\bar{d} : X \longrightarrow \mathbf{GL}(n)$. Hence there is a continuous mapping

$$D : (Y \times I) \cup (X \times \{0\}) \longrightarrow \mathbf{GL}(n).$$

The space $(Y \times I) \cup (X \times \{0\})$ is a retract in the space $X \times I$. Hence D extends to a continuous mapping

$$\bar{D} : X \times I \longrightarrow \mathbf{GL}(n).$$

For nontrivial vector bundles the theorem can be proved by induction with respect to the number of charts. Hence the triple (ξ_1, d_1, ξ_2) is equivalent to a trivial triple if and only there is a trivial triple (η_1, h, η_2) such that the direct sum $(\xi_1 \oplus \eta_1, d \oplus h, \xi_2 \oplus \eta_2)$ is trivial.

What are the inverse elements? We show that the triple $(\xi_1 \oplus \xi_2, d \oplus d^{-1}, \xi_2 \oplus \xi_1)$ is equivalent to trivial triple. For this consider the homotopy

$$D_t = \left\| \begin{matrix} d \sin t & \cos \\ -\cos t & d^{-1} \sin t \end{matrix} \right\| : (\xi_1 \oplus \xi_2)|_Y \longrightarrow (\xi_1 \oplus \xi_2)|_Y$$

When $t = 0$,

$$D_0 = \left\| \begin{matrix} d & 0 \\ 0 & d^{-1} \end{matrix} \right\|$$

and when $t = 1$,

$$D_1 = \left\| \begin{matrix} 0 & 1 \\ -1 & 0 \end{matrix} \right\|.$$

The latter matrix defines an isomorphism which can be extended to an isomorphism over the whole of X. Hence the family $\mathcal{K}^2(X, Y)$ is a group.

Consider a mapping

$$\alpha : K^0(X, Y) \longrightarrow \mathcal{K}^0(X, Y)$$

defined as follows. Let $x \in K^0(X, Y)$ be represented as a difference $x = [\xi] - [\bar{N}]$, $\dim \xi = N$, let $p : X \longrightarrow X/Y$ be the projection, $\eta = p^*(\xi)$, and let $f : \eta \longrightarrow \xi$ be the canonical mapping. Then let

$$d : \eta|_Y \longrightarrow \bar{N}|_Y = Y \times \mathbf{R}^N$$
$$d(h) = (y, f(h)), \; h \in \eta_y, \; y \in Y.$$

The triple $(\eta, d, \bar{N})$, which we denote by $\alpha(x)$, defines an element of the group $\mathcal{K}^0(X, Y)$. This element is well defined, that is, it does not depend on the representation of the element x.

We now prove that α is a monomorphism. Assume that the triple $(\eta, d, \bar{N})$ is equivalent to a trivial triple. This means that there is a trivial triple (η_1, h, η_2) such that direct sum $(\eta \oplus \eta_1, d \oplus h, \bar{N} \oplus \eta_2)$ also is trivial. In particular, the bundles η_1 and η_2 are isomorphic over X. Without loss of generality we can assume that both η_1 and η_2 are trivial and $h \equiv 1$. Hence the triple $(\eta \oplus \bar{N}', d \oplus 1, \bar{N} \oplus \bar{N}')$ is trivial and so the bundle $\eta \oplus \bar{N}'$ is isomorphic to trivial bundle $\bar{N} \oplus \bar{N}'$. Hence $x = 0$.

Next we prove that α is an epimorphism. Consider a triple (ξ_1, d, ξ_2). If η is a bundle such that $\xi_2 \oplus \eta = \bar{N}$. The triple $(\eta, 1, \eta)$ is trivial and the triple $(\xi_1 \oplus \eta, d \oplus 1, \bar{N})$ is equivalent to the triple (ξ_1, d, ξ_2). Hence the bundle $\xi_1 \oplus \eta$ is trivial over Y. This means that $\xi_1 \oplus \eta = p^*(\zeta)$ for a bundle ζ over X/Y. Hence

$$\alpha([\zeta] - [\bar{N}]) = (\xi_1, d, \xi_2).$$

∎

The definition of the groups $\mathcal{K}^0(X, Y)$ has the following generalization. If (ξ_1, d, ξ_2) is a triple, it can be represented as a complex of vector bundles

$$\bar{0} \longrightarrow \xi_1 \overset{\bar{d}}{\longrightarrow} \xi_2 \longrightarrow \bar{0} \tag{3.1}$$

which is exact over Y. For this we need to extend the isomorphism d (over Y) to a morphism $\bar{d}$ of vector bundles over the whole of X. It is clear that such extension exists. Hence a natural generalization of short bundle complexes of type (3.1) is a bundle complex of arbitrary length

$$\bar{0} \longrightarrow \xi_1 \overset{d_1}{\longrightarrow} \xi_2 \overset{d_2}{\longrightarrow} \xi_3 \longrightarrow \cdots \overset{d_{n-1}}{\longrightarrow} \xi_n \longrightarrow \bar{0}, \tag{3.2}$$

which is exact over each point $x \in Y$. There is a natural operation of direct sum of bundle complexes of the type (3.2). The bundle complex (3.2) is said to be *trivial* if it is exact over each point of X. Two bundle complexes of type (3.2) are said to be equivalent if after addition of trivial complexes the corresponding vector bundles become isomorphic and corresponding morphisms become homotopic in the class of morphisms which preserve exactness over Y. The family of classes of equivalent bundle complexes of type (3.2) will be denoted by $\mathcal{K}^n(X, Y)$. There is a natural mapping

$$\beta : \mathcal{K}^2(X, Y) \longrightarrow \mathcal{K}^n(X, Y).$$

A bundle complex of the form

$$\bar{0} \longrightarrow \bar{0} \longrightarrow \cdots \longrightarrow \xi_k \overset{d_k}{\longrightarrow} \xi_{k+1} \longrightarrow \cdots \longrightarrow \bar{0}$$

is said to be an *elementary* complex.

Lemma 6 *Every trivial bundle complex has a splitting into a sum of elementary trivial complexes.*

Proof.

follows from Theorem 7 of the chapter 1.

Lemma 7 *Every bundle complex of type (3.2) is equivalent to an elementary complex, that is, the mapping β is epimorphism.*

Proof.

Consider a bundle complex (3.2) The first step is to change the given complex (3.2) by adding a trivial elementary complex of a sufficiently high dimension in terms $n-2$ and $n-1$:

$$\bar{0}\longrightarrow\xi_1\longrightarrow\cdots\longrightarrow\xi_{n-3}\xrightarrow{D_{n-3}}\xi_{n-2}\oplus\eta\xrightarrow{D_{n-2}}$$
$$\xrightarrow{D_{n-2}}\xi_{n-1}\oplus\eta\xrightarrow{D_{n-1}}\xi_n\longrightarrow\bar{0}, \qquad (3.3)$$

where

$$D_k = d_k, 1\le k\le n-4,$$
$$D_{n-3} = \left\|\begin{matrix}d_{n-3}\\0\end{matrix}\right\|, D_{n-2} = \left\|\begin{matrix}d_{n-2}&0\\0&1\end{matrix}\right\|, D_{n-1} = \|d_{n-1} \quad 0\|.$$

The morphism d_{n-1} is an epimorphism over Y. Hence there is a neighborhood $U\supset Y$ such that d_{n-1} is epimorphism over U. Let φ be a function which equals 1 on $X\backslash U$ and 0 on Y, and φ' be a function which equals 1 on Y and is such that

$$\varphi\varphi'\equiv 0.$$

Consider the new bundle complex

$$\bar{0}\longrightarrow\xi_1\longrightarrow\cdots\longrightarrow\xi_{n-3}\xrightarrow{D'_{n-3}}\xi_{n-2}\oplus\eta\xrightarrow{D'_{n-2}}$$
$$\xrightarrow{D'_{n-2}}\xi_{n-1}\oplus\eta\xrightarrow{D'_{n-1}}\xi_n\longrightarrow\bar{0}, \qquad (3.4)$$

where

$$D'_k = d_k, 1\le k\le n-4,$$
$$D'_{n-3} = \left\|\begin{matrix}d_{n-3}\\0\end{matrix}\right\|, D'_{n-2} = \left\|\begin{matrix}d_{n-2}&0\\0&\varphi'\end{matrix}\right\|, D'_{n-1} = \|d_{n-1} \quad \varphi f\|.$$

and $f:\eta\longrightarrow\xi_n$ is an epimorphism over X.

It is clear that the two complexes (3.3) and (3.4) are homotopic. So we have succeeded in constructing a new equivalent bundle complex (3.4) where the last morphism D_{n-1} is an epimorphism over X. Hence we can split the bundle complex (3.4) into two direct summands:

$$\bar{0}\longrightarrow\xi_1\longrightarrow\cdots\longrightarrow\xi_{n-3}\xrightarrow{D'_{n-3}}\xi_{n-2}\oplus\eta\xrightarrow{D'_{n-2}}\mathbf{Ker}\,(D'_{n-1})\longrightarrow\bar{0},$$

and

$$\bar{0}\longrightarrow \left(\mathbf{Ker}\,(D'_{n-2})\right)^{\perp}\xrightarrow{D'_{n-1}}\xi_n\longrightarrow\bar{0},$$

the latter being a trivial elementary complex. Further, we can apply induction on the length of the bundle complex.

Lemma 8 *The mapping β is a monomorphism.*

The proof is similar to the proof of Lemma 7.

Thus we have shown

Theorem 26 *The natural mapping*

$$\beta : \mathcal{K}^2(X,Y)\longrightarrow\mathcal{K}^n(X,Y).$$

is isomorphism.

Using theorems 25 and 26 we can give different interpretations to important examples of K–groups. The representation of elements of the group $K^0(X,Y)$ as triples or bundle complexes is called the *difference construction*. It is justified by the fact that the restriction of the triple (ξ_1, d, ξ_2) to X equals $[\xi_1] - [\xi_2]$ and the restriction of the bundle complex (3.2) to the space X equals the alternating sum

$$[\xi_1] - [\xi_2] + \ldots + (-1)^{n+1}[\xi_n].$$

Examples

1.   Let us describe the group $K^0(\mathbf{S}^n)$. We represent the sphere $\mathbf{S}^n$ as a quotient space of the disc $\mathbf{D}^n$ by its boundary $\partial\mathbf{D}^n = \mathbf{S}^{n-1}$, $\mathbf{S}^n = \mathbf{D}^n/\mathbf{S}^{n-1}$. Therefore,

$$K^0(\mathbf{S}^n) = \mathcal{K}^2(\mathbf{D}^n, \partial\mathbf{D}^n).$$

Since every bundle over $\mathbf{D}^n$ is trivial, any element of the group $\mathcal{K}^2(\mathbf{D}^n, \partial\mathbf{D}^n)$ may be defined as the homotopy class of a continuous mapping

$$d : \partial\mathbf{D}^n = \mathbf{S}^{n-1}\longrightarrow\mathbf{GL}(N, \mathbf{C}).$$

Hence

$$K^0(\mathbf{S}^2) = \pi_1(\mathbf{U}(N)) = \pi_1(\mathbf{U}(1)) = \pi_1(\mathbf{S}^1) = \mathbf{Z}.$$

Then
$$K(\mathbf{S}^2) = \mathbf{Z} \oplus \mathbf{Z}.$$

Similarly,
$$K^0(\mathbf{S}^3) = \pi_2(\mathbf{U}(N)) = \pi_2(\mathbf{U}(1)) = 0.$$

Hence
$$K(\mathbf{S}^3) = \mathbf{Z}.$$

2. Let SX denote the suspension of the space X, that is, the quotient space of the cone $\mathbf{C}X$ by its base, $SX = \mathbf{C}X/X$. Then

$$K^0(SX) = \mathcal{K}^2(\mathbf{C}X, X) = [X, \mathbf{U}(N)]$$

where $[X, \mathbf{U}(N)]$ is the group of homotopy classes of mappings of X to unitary group $\mathbf{U}(N)$, for sufficiently large N.

3. Consider a generalization of the difference construction. Let us substitute a triple (ξ_1, d, ξ_2) for two complexes of type (3.2) and a homomorphism giving a commutative diagram

$$
\begin{array}{ccccccccc}
0 & \longrightarrow & \xi_1 & \xrightarrow{d_{11}} & \xi_2 & \xrightarrow{d_{12}} & \cdots & \xrightarrow{d_{1,n-1}} & \xi_n & \longrightarrow & 0 \\
 & & \downarrow{\scriptstyle d'_1} & & \downarrow{\scriptstyle d'_2} & & & & \downarrow{\scriptstyle d'_n} & & \\
0 & \longrightarrow & \eta_1 & \xrightarrow{d_{21}} & \eta_2 & \xrightarrow{d_{22}} & \cdots & \xrightarrow{d_{2,n-1}} & \eta_n & \longrightarrow & 0
\end{array}
\qquad (3.5)
$$

Assume that horizontal homomorphisms d_{ij} are exact over Y and vertical homomorphisms d'_k are exact over Y'. It is natural to expect that the diagram (3.5) defines an element of the group $K^0(X/(Y \cup Y'))$. Indeed, there is a new complex

$$0 \longrightarrow \xi_1 \xrightarrow{D_1} (\xi_2 \otimes \eta_1) \xrightarrow{D_2} (\xi_3 \otimes \eta_2) \longrightarrow \cdots \longrightarrow (\xi_n \otimes \eta_{n-1}) \longrightarrow \eta_n \longrightarrow 0, \qquad (3.6)$$

in which the horizontal homomorphisms are defined by the matrices

$$D_k = \left\| \begin{array}{cc} d_{1k} & 0 \\ d'_k & d_{2,k-1} \end{array} \right\|.$$

This complex (3.6) is exact over the subspace $(Y \cup Y')$.

More generally, consider a diagram

$$
\begin{array}{ccccccccc}
0 & \longrightarrow & \xi_{11} & \xrightarrow{d_{11}} & \xi_{12} & \xrightarrow{d_{12}} & \dots & \xrightarrow{d_{1,n-1}} & \xi_{1n} & \longrightarrow & 0 \\
& & \downarrow{\scriptstyle d'_{11}} & & \downarrow{\scriptstyle d'_{12}} & & & & \downarrow{\scriptstyle d'_{1n}} & & \\
0 & \longrightarrow & \xi_{21} & \xrightarrow{d_{21}} & \xi_{22} & \xrightarrow{d_{22}} & \dots & \xrightarrow{d_{2,n-1}} & \xi_{2n} & \longrightarrow & 0 \\
& & \downarrow & & \downarrow & & & & \downarrow & & \\
& & \vdots & & \vdots & & & & \vdots & & \\
& & \downarrow & & \downarrow & & & & \downarrow & & \\
0 & \longrightarrow & \xi_{m1} & \xrightarrow{d_{m1}} & \xi_{m2} & \xrightarrow{d_{m2}} & \dots & \xrightarrow{d_{m,n-1}} & \xi_{mn} & \longrightarrow & 0 \\
& & \downarrow & & \downarrow & & & & \downarrow & & \\
& & 0 & & 0 & & & & 0 & &
\end{array}
$$

with anticommutative squares:

$$d'_{k,l+1}d_{kl} + d_{k+1,l}d'_{kl} = 0.$$

Assume that the horizontal rows are exact over a subset Y and the vertical columns are exact over a subset Y'. Then this diagram defines an element of the group $K^0(X, Y \cup Y')$.

4. The previous construction allows us to construct a natural tensor product of K-groups as a bilinear function

$$K^0(X,Y) \times K^0(X',Y') \longrightarrow K^0(X \times X', X \times Y' \cup Y \times Y'). \qquad (3.7)$$

Consider a complex on X,

$$0 \longrightarrow \xi_1 \xrightarrow{d_1} \xi_2 \xrightarrow{d_2} \dots \xrightarrow{d_{n-1}} \xi_n \longrightarrow 0 \qquad (3.8)$$

which is exact over Y and a complex on X'

$$0 \longrightarrow \eta_1 \xrightarrow{d_1} \eta_2 \xrightarrow{d_2} \dots \xrightarrow{d_{n-1}} \eta_n \longrightarrow 0 \qquad (3.9)$$

which is exact over Y'. Consider the tensor product of the two complexes (3.8) and (3.9)

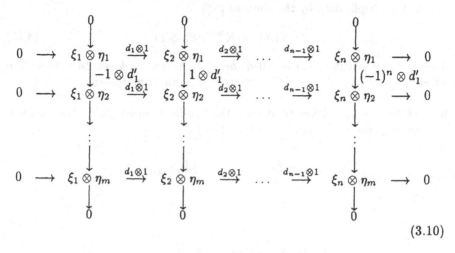

$$(3.10)$$

It is easy to check that this diagram has rows exact over $X \times Y'$ and columns exact over $Y \times X'$. Hence the diagram (3.10) defines a tensor product as a bilinear operation of type (3.7)

If $X = X'$ then using diagonal inclusion $X \subset X \times X$ we have the tensor product as a bilinear operation

$$K(X, Y) \times K(X, Y') \longrightarrow K(X, Y \cup Y').$$

5. Consider a complex n–dimensional vector bundle ξ over the base X and let $p : E \longrightarrow X$ be the projection of the total space E onto the base X. Consider the space E as a new base space and a complex of vector bundles

$$0 \longrightarrow \Lambda_0 \eta \xrightarrow{d_0} \Lambda_1 \eta \xrightarrow{d_1} \Lambda_2 \eta \xrightarrow{d_2} \ldots \xrightarrow{d_{n-1}} \Lambda_n \eta \longrightarrow 0, \qquad (3.11)$$

where $\eta = p^*\xi$ is inverse image of the bundle ξ and the homomorphism

$$d_k : \Lambda_k \eta \longrightarrow \Lambda_{k+1} \eta$$

is defined as the exterior multiplication by the vector $y \in E$, $y \in \xi_x$, $x = p(y)$. It is known that if the vector $y \in \xi_x$ is nonzero, $y \neq 0$, then the complex (3.11) is exact. Consider the subspace $D(\xi) \subset E$ which consists of all vectors $y \in E$ such that $|y| \leq 1$. Then the subspace $S(\xi) \subset D(\xi)$ of all unit vectors gives the pair

$(D\xi), S(\xi))$ for which the complex (3.11) is exact on $S(\xi)$. Denote the element defined by (3.11) by $\beta(\xi) \in K^0(D(\xi), S(\xi))$. Then one has homomorphism given by multiplication by the element $\beta(\xi)$

$$\beta : K(X) \longrightarrow K^0(D(\xi), S(\xi)). \qquad (3.12)$$

The homomorphism (3.12) is called *the Bott homomorphism*. In next section we shall prove that the Bott homomorphism is an isomorphism.

6. There is a simple way to shorten the bundle complex (3.2). Without loss of generality we may assume that $n = 2m$. Put

$$\eta_0 = \xi_1 \oplus \xi_3 \oplus \ldots = \bigoplus_{k=1}^{m} \xi_{2k-1},$$

$$\eta_1 = \xi_2 \oplus \xi_4 \oplus \ldots = \bigoplus_{k=1}^{m} \xi_{2k},$$

$$D = \begin{Vmatrix} d_1 & d_2^* & 0 & \cdots & 0 \\ 0 & d_3 & d_4^* & \cdots & 0 \\ 0 & 0 & d_5 & \cdots & 0 \\ & & \cdots & & \\ 0 & 0 & 0 & \cdots & d_{2m-2}^* \\ 0 & 0 & 0 & \cdots & d_{2m-1} \end{Vmatrix}.$$

First of all notice that there is a sequence

$$\cdots \longrightarrow \eta_0 \xrightarrow{D_0} \eta_1 \xrightarrow{D_1} \eta_0 \xrightarrow{D_0} \cdots,$$

where

$$D_0 = \begin{Vmatrix} d_1 & 0 & \cdots & 0 \\ 0 & d_3 & \cdots & 0 \\ & & \cdots & \\ 0 & 0 & \cdots & d_{2m-1} \end{Vmatrix}, \quad D_1 = \begin{Vmatrix} 0 & 0 & \cdots & 0 & 0 \\ d_2 & 0 & \cdots & 0 & 0 \\ & & \cdots & & \\ 0 & 0 & \cdots & d_{2m-2} & 0 \end{Vmatrix}.$$

This sequence is exact over Y. On the other hand, $D = D_0 + D_1^*$. Hence D is isomorphism over Y.

7. Using the previous construction we can define the tensor product of two triples

$$0\longrightarrow\xi_1\xrightarrow{d}\xi_2\longrightarrow0,\ 0\longrightarrow\eta_1\xrightarrow{d'}\eta_2\longrightarrow0$$

as a triple

$$0\longrightarrow(\xi_1\otimes\eta_1)\oplus(\xi_2\otimes\eta_2)\xrightarrow{D}(\xi_1\otimes\eta_2)\oplus(\xi_2\otimes\eta_1)\longrightarrow0,$$

where

$$D=\begin{Vmatrix}d\otimes1 & -1\otimes d'^*\\1\otimes d' & d^*\otimes1\end{Vmatrix}.$$

3.2 BOTT PERIODICITY

Consider the Bott element $\beta=\beta(1)\in K^0\left(\mathbf{S}^2\right)$ which was defined in the section 3.1. The two natural projections from $X\times\mathbf{S}^2$ to the factors X and $\mathbf{S}^2$ generate a natural homomorphisms from $K(X)$ and $K\left(\mathbf{S}^*\right)$ to the ring $K\left(X\times\mathbf{S}^*\right)$. Then since $K\left(\mathbf{S}^2\right)\sim\mathbf{Z}[\beta]/\{\beta^2=0\}$, there is a homomorphism

$$h:K(X)[t]/\{t^2=0\}\longrightarrow K\left(X\times\mathbf{S}^2\right),\qquad(3.13)$$

defined by the formula

$$h(xt+y)=x\beta+y.$$

Theorem 27 (Bott periodicity) *The homomorphism (3.13) is an isomorphism.*

Proof.

We need to construct an inverse homomorphism to (3.13). Consider the pair $\left(X\times\mathbf{S}^2,X\times\{s_0\}\right)$, where s_0 is a fixed point. From the exact sequence in K–theory:

$$K\left(S\left(X\times\mathbf{S}^2\right)\right)\xrightarrow{Si^*}K(SX)\xrightarrow{\delta}K\left(X\times\mathbf{S}^2,X\times\{s_0\}\right)\xrightarrow{j^*}K\left(X\times\mathbf{S}^2\right)\xrightarrow{i^*}K(X)$$

we see that δ is trivial since i^* and Si^* are epimorphisms. Hence the homomorphism h sends the element $ty\in K(X)[t]/\{t^2=0\}$ to $K\left(X\times\mathbf{S}^2,X\times\{s_0\}\right)$. Denote the restriction of this homomorphism by $\tilde{h}$:

$$\tilde{h}:K(X)\longrightarrow K\left(X\times\mathbf{S}^2,X\times\{s_0\}\right).\qquad(3.14)$$

$$\tilde{h}(y) = \beta y, \; y \in K(X).$$

The pair $\left(X \times \mathbf{S}^2, X \times \{s_0\}\right)$ is equivalent to the pair $\left(X \times \mathbf{D}^2, X \times \mathbf{S}^1\right)$. Hence each element of the group $K^0\left(X \times \mathbf{D}^2, X \times \mathbf{S}^1\right)$ is represented by a triple (ξ_1, d, ξ_2) where d is an isomorphism over $X \times \mathbf{S}^1$. Moreover, we can assume that both vector bundles ξ_1 and ξ_2 are trivial. Thus the isomorphism d is a continuous function

$$d : X \times \mathbf{S}^1 \longrightarrow \mathbf{U}(n)$$

for sufficiently large n. Equivalent triples can be obtained by using the two operations: stabilization of dimension generated by inclusion $\mathbf{U}(n) \subset \mathbf{U}(n+n')$ and homotopies of d in the class of continuous mappings $X \times \mathbf{S}^1 \longrightarrow \mathbf{GL}(\mathbf{C}, n)$.

Thus we have a continuous mapping

$$d(x, z) \in \mathbf{U}(n), \; x \in X, \; z \in \mathbf{S}^1 \subset \mathbf{C}, \; |z| = 1.$$

The first step

We produce a smooth ε–approximation of the function $d(x, z)$, for example, by the formula

$$d_1(x, z) = \frac{q}{2\varepsilon} \int\limits_{\varphi-\varepsilon}^{\varphi+\varepsilon} d(x, e^{i\varphi}) d\varphi.$$

The second step

Consider the Fourier series:

$$d_1(x, z) = \sum_{k=-\infty}^{\infty} a_k(x) z^k, \tag{3.15}$$

where

$$a_k(x) = \frac{1}{2\pi} \int\limits_{0}^{2\pi} d_1(x, e^{-ik\varphi}) d\varphi.$$

The series (3.15) converges uniformly with respect to the variables x and z and equipotentially in the class of continuous functions $d_1(x, z) \in \mathbf{U}(n)$.

The third step

Restrict to a finite part of the Fourier series:

$$f_N(x, z) = \sum_{k=-N}^{N} a_k(x) z^k.$$

The number N can be chosen sufficiently large so that

$$f_N(x, z) \in \mathbf{GL}(\mathbf{C}, n).$$

The fourth step

Put

$$p_N(x, z) = z^N f_N(x, z).$$

The fifth step

Let

$$p_N(x, z) = \sum_{k=0}^{2N} b_k(x) z^k \in \mathbf{GL}(\mathbf{C}, n).$$

Put

$$\mathcal{L}_N(p)(x, z) = \begin{Vmatrix} b_0(x) & b_1(x) & \ldots & b_{N-1}(x) & b_N(x) \\ -z & 1 & \ldots & 0 & 0 \\ 0 & -z & \ldots & 0 & 0 \\ \vdots & \vdots & & \vdots & \vdots \\ 0 & 0 & \ldots & -z & 1 \end{Vmatrix} \in \mathbf{GL}(\mathbf{C}, nN). \quad (3.16)$$

The sixth step

The function (3.16) is linear with respect to the variable z,

$$\mathcal{L}_N(p)(x, z) = A(x) z + B(x) \in \mathbf{GL}(\mathbf{C}, nN).$$

Hence there is a projection

$$Q(x) : \mathbf{C}^{nN} \longrightarrow \mathbf{C}^{nN}$$

which commutes with $\mathcal{L}_N(p)(x,z)$ and satisfies the following condition: the space $\mathbf{C}^{nN}$ is split into two summands

$$\mathbf{C}^{nN} = V_+(x) \oplus V_-(x)$$

such that the operator $\mathcal{L}_N(p)(x,z)$ is an isomorphism on $V_+(x)$ when $|z| \geq 1$ and is an isomorphism on $V_-(x)$ when $|z| \leq 1$.

The seventh and final step

The family of $V_+(x)$ induces a vector subbundle of the trivial bundle $X \times \mathbf{C}^{nN}$. Put

$$\nu_N(d) = ([V_+(x)] - \overline{nN}) . \tag{3.17}$$

We need to prove that the definition (3.17) gives the homomorphism inverse to (3.14).

3.3   PERIODIC K–THEORY

Using the Bott periodicity we can define the groups $K^n(X,Y)$ for any integer n, negative or positive. First of all notice that the direct sum $\bigoplus\limits_{n \leq 0} K^n(X,Y)$ can be given a ring structure. In fact,

$$K^{-n}(X,Y) = K^0\left(X \times \mathbf{D}^n, \left((Y \times \mathbf{D}^n) \cup (X \times \mathbf{S}^{n-1})\right)\right) .$$

Hence the operation of tensor product considered in the example 3.1 of the section 3.1 gives the following pairing

$$K^{-n}(X,Y) \times K^{-m}(X',Y') =$$

$$\begin{aligned}
= \quad & K^0\left(X \times \mathbf{D}^n, \left((Y \times \mathbf{D}^n) \cup (X \times \mathbf{S}^{n-1})\right)\right) \times \\
& \qquad \times K^0\left(X' \times \mathbf{D}^m, \left((Y' \times \mathbf{D}^m) \cup (X' \times \mathbf{S}^{m-1})\right)\right) \longrightarrow \\
\longrightarrow \quad & K^0(X \times X' \times \mathbf{D}^n \times \mathbf{D}^m, (X \times Y' \times \mathbf{D}^n \times \mathbf{D}^m) \cup \\
& \qquad \cup (X \times X' \times \mathbf{D}^n \times \mathbf{S}^{m-1}) \cup \\
& \cup (Y \times X' \times \mathbf{D}^n \times \mathbf{D}^m) \cup (X \times X' \times \mathbf{S}^{n-1} \times \mathbf{D}^m)) = \\
= \quad & K^0(X \times X' \times \mathbf{D}^n \times \mathbf{D}^m, (((X \times Y') \cup (Y \times X')) \times \mathbf{D}^n \times \mathbf{D}^m) \cup
\end{aligned}$$

$$\cup (X \times X' \times \mathbf{D}^n \times \mathbf{S}^{m-1}) \cup (X \times X' \times \mathbf{S}^{n-1} \times \mathbf{D}^m)) =$$
$$= K^0 (X \times X' \times \mathbf{D}^{n+m},$$
$$((Y \times X') \cup (X \times Y') \times \mathbf{D}^{n+m}) \cup (X \times X' \times \mathbf{S}^{n+m-1})) =$$

$$= K^{-(n+m)}(X \times X', (X \times Y') \cup (Y \times X')).$$

In particular, when $X = X'$ and $Y = Y'$ it gives the usual inner multiplication

$$K^{-n}(X, Y) \times K^{-m}(X, Y) \longrightarrow K^{-(n+m)}(X, Y).$$

When $X' = \mathbf{S}^0$, $Y' = \{s_0\}$, where $\mathbf{S}^0$ is a 0–dimensional sphere, we have the pairing

$$K^{-n}(X, Y) \times K^{-m}(\mathbf{S}^0, s_0) \longrightarrow$$
$$\longrightarrow K^{-(n+m)}(X \times \mathbf{S}^0, (X \times s_0) \cup (Y \times \mathbf{S}^0)) =$$
$$= K^{-(n+m)}(X, Y).$$

The Bott element $\beta \in K^0(\mathbf{S}^2, s_0) = K^{-2}(\mathbf{S}^0, s_0) = \mathbf{Z}$ generates a homomorphism $\tilde{h}$:

$$K^{-n}(X, Y) \xrightarrow{\otimes \beta} K^{-(n+2)}(X, Y) \qquad (3.18)$$

by the Bott periodicity isomorphism. This formula (3.18) justifies the term 'Bott periodicity'.

Thus, via the isomorphism (3.18), the groups $K^n(X, Y)$ can be indexed by the integers modulo 2. Then the exact sequence for the pair (X, Y) has the following form :

$$\cdots \quad \longrightarrow K^0(X, Y) \longrightarrow K^0(X, x_0) \longrightarrow K^0(Y, x_0) \longrightarrow$$
$$\longrightarrow K^1(X, Y) \longrightarrow K^1(X, x_0) \longrightarrow K^1(Y, x_0) \longrightarrow$$
$$\longrightarrow K^0(X, Y) \longrightarrow \quad \cdots \qquad (3.19)$$

Let

$$K^*(X, Y) = K^0(X, Y) \oplus K^1(X, Y).$$

Then the sequence (3.19) can be written as

$$\cdots \quad \longrightarrow K^*(X, Y) \xrightarrow{j^*} K^*(X, x_0) \xrightarrow{i^*} K^*(Y, x_0) \xrightarrow{\partial}$$
$$\xrightarrow{\partial} K^*(X, Y) \xrightarrow{j^*} K^*(X, x_0) \xrightarrow{i^*} K^*(Y, x_0) \xrightarrow{\partial}$$
$$\xrightarrow{\partial} K^*(X, Y) \longrightarrow \quad \cdots$$

where i^* and j^* are ring homomorphisms and ∂ is an operator of degree 1.

Consider the Chern character defined in the section 2.6. It can be represented as a ring homomorphism of graded rings

$$\mathrm{ch} : K^*(X,Y) \longrightarrow H^{**}(X,Y;\ \mathbf{Q}),$$

where $H^{**}(X,Y;\ \mathbf{Q})$ is considered as a $\mathbf{Z}_2$-graded ring by decomposition into a direct sum of the even and odd dimensional cohomology.

Theorem 28 *The homomorphism*

$$\mathrm{ch} : K^*(X,Y) \otimes \mathbf{Q} \longrightarrow H^{**}(X,Y;\ \mathbf{Q})$$

is an isomorphism for any finite cellular pair (X,Y).

Proof.

The proof is based on the Five Lemma and uses induction on the number of cells. The initial step consists of checking that the homomorphism

$$\mathrm{ch} : K^*(\mathbf{D}^n, \mathbf{S}^{n-1}) \otimes \mathbf{Q} \longrightarrow H^{**}(\mathbf{D}^n, \mathbf{S}^{n-1};\ \mathbf{Q})$$

is an isomorphism.

Let $n = 2$. Then the

$$K^0(\mathbf{D}^2, \mathbf{S}^1) \sim \mathbf{Z}$$

with generator

$$\beta = [\xi] - \mathbf{1},$$

where ξ is the Hopf bundle. Then

$$\mathrm{ch}\,\beta = \mathrm{ch}\,\xi - 1 = e^{c_1} - 1,$$

where

$$c_1 \in H^2(\mathbf{S}^2, s_0;\ \mathbf{Q})$$

is the integer generator. Hence the homomorphism

$$\mathrm{ch} : K^0(\mathbf{D}^2, \mathbf{S}^1) \otimes \mathbf{Q} \longrightarrow H^2(\mathbf{D}^2, \mathbf{S}^1;\ \mathbf{Q})$$

is an isomorphism. For the odd case, one has

$$K^1(\mathbf{D}^2, \mathbf{S}^1) = H^{\mathrm{odd}}(\mathbf{D}^2, \mathbf{S}^1) = 0.$$

Thus we have proved initial step. This finishes the proof of theorem.

Corollary 5 *For any element $x \in H^{2k}(X)$ there is a bundle ξ and a number λ such that*

$$ch\,\xi = \dim\xi + \lambda x + \text{ terms of higher dimension}.$$

3.4 LINEAR REPRESENTATIONS AND BUNDLES

Consider a linear representation of a group G on a (complex) n–dimensional vector space

$$\rho : \mathbf{G}\longrightarrow\mathbf{GL}(n,\mathbf{C}).$$

If ξ is a principal G–bundle over a base space X with transition functions

$$\varphi_{\alpha\beta} : U_{\alpha\beta}\longrightarrow G$$

then using the new transition functions

$$\psi_{\alpha\beta} = \rho \circ \varphi_{\alpha\beta} : U_{\alpha\beta}\longrightarrow\mathbf{GL}(n,\mathbf{R})$$

we obtains a new vector bundle ξ^ρ.

We apply this construction to the universal principal G–bundle ξ_G over the classifying space B_G. Denote by $R(G)$ the group of virtual finite dimensional representations of the group G. It is clear that the group $R(G)$ is a ring with respect to tensor product of representations. By the previous construction there is a ring homomorphism

$$b : R(G)\longrightarrow K(BG). \tag{3.20}$$

Interest in the homomorphism (3.20) comes from the fact that it has applications to a number of different problems. The virtual bundle $b(\rho) = [\xi_G^\rho]$ is an invariant of the virtual representation ρ. This bundle can be used to interpret algebraic properties of the representation ρ in homotopy terms.

Let $G = \mathbf{U}(n)$. Then elements of the group $K(\mathbf{BU}(n))$ correspond to cohomology operations in K–theory. If an element α belongs to the image of (3.20) then the corresponding cohomology operation is determined by a linear representation of the structure group $\mathbf{U}(n)$. In particular, it could be interest to know if the homomorphism (3.20) is an epimorphism or to describe the image of (3.20).

Consider some simple examples.

The group $G = \mathbf{S}^1$.

Each linear complex representation of the group $\mathbf{S}^1 = \mathbf{U}(1)$ can be split into a direct sum of one–dimensional representations which in turn are described by the following formula

$$\rho_k(z) = z^k,$$

where $z \in G = \mathbf{S}^1$ is a complex number with unit modulus. So the irreducible representations are parametrized by the integers. It is clear that

$$\rho_k \otimes \rho_s = \rho_{k+s}.$$

Hence the ring $R(\mathbf{S}^1)$ is generated by two elements ρ_1 and ρ_{-1} with the unique relation $\rho_1 \rho_{-1} = 1$. Thus,

$$R(\mathbf{S}^1) = \mathbf{Z}[\rho_1, \rho_{-1}]/\{\rho_1 \rho_{-1} = 1\},$$

the so called *ring of Laurent polynomials*.

On the other hand, the classifying space BG is homeomorphic to the direct limit

$$BG = \lim_{n \to \infty} \mathbf{CP}^n = \mathbf{CP}^\infty.$$

The representation ρ_1 is the identity representation. Hence

$$b(\rho_1) = \xi_1$$

and

$$b(\rho_k) = \underbrace{\xi_1 \otimes \ldots \otimes \xi_1}_{k \text{ times}}$$

for $k \geq 0$. Notice that the representation ρ_{-1} is the complex conjugate of ρ_1 since $z^{-1} - \bar{z}$ for $|z| = 1$. Hence

$$b(\rho_{-k}) = \underbrace{\bar{\xi}_1 \otimes \ldots \otimes \bar{\xi}_1}_{k \text{ times}}$$

Thus

Theorem 29 *The image of the homomorphism*

$$b : R(\mathbf{S}^1) \otimes \mathbf{Q} \longrightarrow K(\mathbf{BS}^1) \otimes \mathbf{Q},$$

where $K(\mathbf{BS}^1) \otimes \mathbf{Q}$ is the completion with respect to the topology generated by the ideal of zero dimensional virtual bundles.

Proof.

Consider the homomorphism

$$R(\mathbf{S}^1) \xrightarrow{b} K(\mathbf{BS}^1) \xrightarrow{\text{ch}} H^{**}(\mathbf{BS}^1; \mathbf{Q})$$

After tensoring with $\mathbf{Q}$ we have

$$R(\mathbf{S}^1) \otimes \mathbf{Q} \xrightarrow{b} K(\mathbf{BS}^1) \otimes \mathbf{Q} \xrightarrow{\text{ch}} H^{**}(\mathbf{BS}^1; \mathbf{Q}) \qquad (3.21)$$

The group $K(\mathbf{BS}^1)$ is defined to be the inverse limit

$$K(\mathbf{BS}^1) = \varprojlim K(\mathbf{CP}^n).$$

Since

$$H^{**}(\mathbf{BS}^1; \mathbf{Q}) = \varprojlim H^{**}(\mathbf{CP}^n; \mathbf{Q}),$$

using theorem 28 we have that the homomorphism ch in (3.21) is a topological isomorphism. On the other hand, for $u = \rho_1 - 1$,

$$\text{ch } b(u) = a + \frac{a^2}{2} + \cdots + \frac{a^k}{k!} + \cdots.$$

Hence this element can serve as a generator in the ring of formal series $H^{**}(\mathbf{BS}^1; \mathbf{Q})$.

Consider the ideal of zero–dimensional virtual representations in $R(G)$. The corresponding topology generates the completion $\widehat{R(G)}$ of the ring $R(G)$. It is easy to prove that homomorphism

$$b : \widehat{R(\mathbf{S}^1)} \otimes \mathbf{Q} \longrightarrow K(\mathbf{BS}^1) \otimes \mathbf{Q}$$

is an isomorphism. ∎

The group $G = \mathbf{Z}$.

The classifying space $B\mathbf{Z}$ can be taken to be homeomorphic to the circle $\mathbf{S}^1$ and $K(\mathbf{S}^1) = \mathbf{Z}$. Hence linear representations of the group $\mathbf{Z}$ are not distinguished by the homomorphism b.

The group $G = \mathbf{Z} \times \mathbf{Z}$.

In this case the classifying space can be taken to be homeomorphic to $T^2 = \mathbf{S}^1 \times \mathbf{S}^1$. The group $K(T^2)$ can be calculated using the exact sequence of the pair $(T^2, \mathbf{S}^1 \vee \mathbf{S}^1)$. Finally,

$$K^0(T^2, s_0) \;=\; \mathbf{Z},$$
$$K^1(T^2, s_0) \;=\; \mathbf{Z} \oplus \mathbf{Z}.$$

Each irreducible representation ρ of the commutative group $\mathbf{Z} \times \mathbf{Z}$ is one–dimensional and homotopic to the trivial representation. Thus $b(\rho)$ is trivial.

The group $G = \mathbf{Z}^n$ and the case of a family of representations.

The case of $G = \mathbf{Z}^n$ is similar to the case of $G = \mathbf{Z}^2$. But one can obtain a nontrivial result by considering a continuous family of representations. Namely, let ρ_y, $y \in Y$ be a continuous family of representations. Let ξ_G be the universal principal G–bundle over $X = BG$ with the transition functions $\varphi_{\alpha\beta}$. We can consider the same transition functions over the base $X \times Y$ by taking them independent of the argument $y \in Y$. The compositions $\psi_{\alpha\beta} = \rho_y \circ \varphi_{\alpha\beta}$ depend on both arguments $x \in X, y \in Y$. The latter defines a vector bundle ξ_G^ρ over the base $X \times Y$. Then in spite of the fact that the restriction of vector bundle $\xi_G^{\rho_{y_0}}$ over each $X \times y_0$ may be trivial the whole vector bundle ξ_G^ρ over $X \times Y$ may be nontrivial.

For example, consider as the space Y, the character group of the group G,

$$Y = G^*,$$

and consider as representation ρ_y the corresponding character of the group G,

$$\rho_y : G \longrightarrow \mathbf{S}^1 = \mathbf{U}(1).$$

When $G = \mathbf{Z}^n = \bigoplus G_j$, $G_j = \mathbf{Z}$,

$$G^* = T^n \text{ and } BG = T^n.$$

Then

$$X \times Y = BG \times G^* = T^n \times T^n.$$

Theorem 30 *Let ρ be the family of all characters of the group $G = \mathbf{Z}^n$. Then*

$$c_1(\xi_G^\rho) = x_1 y_1 + \ldots + x_n y_n$$

where $x_k \in H^1(BG; \mathbf{Z})$, $y_k \in H^1(G^*; \mathbf{Z})$ are corresponding generators.

Proof.

The two dimensional cohomology of the space $X \times Y$ is generated by monomials of degree 2 in the variables x_k, y_k. Hence

$$c_1(\xi_G^\rho) = \sum \lambda_{ij} x_i y_j + \sum \mu_{ij} x_i x_j + \sum \nu_{ij} y_i y_j$$

for some integers $\lambda_{ij}, \mu_{ij}, \nu_{ij}$. It is clear that the second and the third sums are trivial by considering the characteristic classes of restrictions of the bundles over the factors. The coefficients λ_{ij} can be interpreted as characteristic classes of restrictions to the subspaces $BG_i \times G_j^*$ Therefore, it is clear that $\lambda_{ij} = 0$ when $i \neq j$. Hence the proof reduces to the case where $G = \mathbf{Z}$. In this case $X = \mathbf{S}^1$ and $Y = \mathbf{S}^1$. There is an atlas consisting of two charts: the upper and lower semicircles, U_1 and U_2. The intersection $U_{12} = U_1 \cap U_2$ consists of two points: $\{1, -1\}$. The principal bundle ξ_G has one transition function φ_{12} where

$$\varphi_{12}(1) = 0 \in \mathbf{Z}, \quad \varphi_{12}(-1) = 1 \in \mathbf{Z}.$$

Let $y \in Y = \mathbf{S}^1$ be a character of the group $G = \mathbf{Z}$. Then the corresponding transition function $\psi_{12} = \rho_y \circ \varphi_{12}$ is defined by the formula

$$\psi_{12}(1, y) = 1, \quad \psi_{12}(-1, y) = y.$$

It is easy to see that the bundle ξ_G^ρ is trivial over the wedge of a parallel and a meridian $\mathbf{S}^1 \vee \mathbf{S}^1$ and on the quotient space $T^2/(\mathbf{S}^1 \vee \mathbf{S}^1) = \mathbf{S}^2$ it is isomorphic to the Hopf bundle. Hence

$$c_1(\xi_G^\rho) = xy \in H^2(T^2; \mathbf{Z}).$$

∎

3.5 EQUIVARIANT BUNDLES

Let G be a compact Lie group. A *G–space* X is a topological space X with continuous action of the group G on it. The map $f : X \longrightarrow Y$ is said to be *equivariant* if

$$f(gx) = gf(x), g \in G.$$

Similarly, if f is a locally trivial bundle and also equivariant then f is called an *equivariant* locally trivial bundle. An equivariant vector bundle is defined similarly. The theory of equivariant vector bundles is very similar to the classical theory. In particular, equivariant vector bundles admit the operations of direct sum and tensor product. In certain simple cases the description of equivariant vector bundles is reduced to the description of the usual vector bundles.

Consider first the case where the action of the group G is *free*, then the natural projection π onto the orbit space $Y = X/G$ gives a locally trivial bundle.

Theorem 31 *The family of G-equivariant vector bundles over a space X with a free action of a group G is in one-to-one correspondence with the family of vector bundles over the orbit space $Y = X/G$. This correspondence is given by the projection π:*

$$\eta \longrightarrow \pi^* \eta$$

with the natural action of the group G on the total space of vector bundle ψ.

Proof.

In fact, in the total space of the bundle $\pi^*(\eta)$, there is natural action of the group G. By definition the charts of the bundle $\pi^*(\eta)$ are inverse images of the charts U on Y. Then

$$\pi^{-1}(U) \sim U \times G$$

with a left action along the second summand. The corresponding trivialization of the vector bundle $\pi^*(\eta)$ is homeomorphic to $U \times G \times F$ where F is the fiber. Then, by definition, the action of the group G on the total space is the left action along the second summand.

Conversely, let $\xi \longrightarrow X$ be arbitrary G-equivariant vector bundle with a free G action. Consider a small chart $U \subset Y$ and put $V = \pi^{-1}(U) = U \times G$. For U sufficiently small that the restriction of ξ to $U \times g_0$ is trivial

$$\xi_{U \times g_0} \sim U \times g_0 \times F.$$

Then the restriction of the bundle $\xi_{U \times G}$ also is trivial:

$$U \times g_0 \times G \times F \quad \longrightarrow \quad \xi_{U \times G},$$
$$(u, g_0, g, f) \longrightarrow g(u, g_0, f).$$

It is clear that the transition functions do not depend on the argument $g \in G$.

∎

The second case occurs when the group G acts trivially on X.

Theorem 32 *Suppose the group G acts trivially on the base X. Then for each G–equivariant vector bundle x there exist irreducible representations of the group G in vector spaces V_i and vector bundles ξ_i such that*

$$\xi = \bigoplus_i (\xi_i \otimes V_i)$$

with the natural action of the group G on the second factors.

Proof.

The action of the group G preserves each fiber and acts there as a linear representation. Hence this representation can be split into a direct sum of irreducible representations. This means that the fiber ξ_x gives a direct sum

$$\xi_x = \oplus_i (\xi_{x,i} \otimes V_i). \tag{3.22}$$

Assume that locally the presentation (3.22) does not depend on the point x, that is, in some small neighborhood U_α we have

$$\xi_{|U_\alpha} = \xi_{x_0} \times U_\alpha = [\oplus_i (\xi_{x_0,i} \otimes V_i)] \times U_\alpha. \tag{3.23}$$

Then using Schur's lemma the equivariant transition function $\psi_{\alpha\beta}$ decomposes into a direct sum

$$\psi_{\alpha\beta} = \oplus_i (\psi_{\alpha\beta,i} \otimes V_i).$$

The functions $\psi_{\alpha\beta,i}$ define transition functions for some vector bundles ξ_i. Thus

$$\xi = \oplus_i (\xi_i \otimes V_i).$$

Our problem is to prove the decomposition (3.23). We give here an elegant proof due to M.F. Atiyah. Consider a G–equivariant vector bundle $E \longrightarrow X$ with the trivial action on the base X. Denote by E^G the set of fixed points. Then $E^G \longrightarrow X$ also is a locally trivial vector bundle. To prove this it is sufficient to construct a fiberwise projection P onto the set of fixed points. Put

$$P(v) = \int_G g(v) d\mu(g)$$

where $d\mu(g)$ is the Haar measure on the compact group G. It is clear that P is a projection since

$$P(P(v)) = \int_G h(P(v)) d\mu(h) = \iint_{G \times G} hg(v) d\mu(h) d\mu(g) =$$

$$= \iint\limits_{G \times G} g'(v) d\mu(g') d\mu(h') = \int\limits_G g'(v) d\mu(g') \cdot \int\limits_G d\mu(h') =$$

$$= \int\limits_G g'(v) d\mu(g') = P(v).$$

On the other hand,
$$g(P(v)) = P(v),$$
and if $g(v) = v$ then $P(v) = v$. Hence the image of P coincides with E^G. Using Theorem 7 it can be shown that E^G is a locally trivial bundle.

Consider the bundle $\mathbf{HOM}\,(\xi_1, \xi_2)$ (for the definition see section 1.3) . If both bundles ξ_1 and ξ_2 are G–equivariant then the bundle $\mathbf{HOM}\,(\xi_1, \xi_2)$ also is G–equivariant with the action defined as follows:

$$\varphi \longrightarrow g\varphi g^{-1}.$$

Then the fixed point set $\mathbf{HOM}\,_G(\xi_1, \xi_2)$ gives a locally trivial bundle. Now we can construct the bundles ξ_i mentioned above. Let V_i be an irreducible G–vector space and $\underline{V_i}$ be the corresponding trivial G–equivariant vector bundle. Put
$$\xi_i = \mathbf{HOM}\,_G(\underline{V_i}, \xi).$$
Then there is an isomorphism

$$\alpha : \oplus_i \left(\underline{V_i} \otimes \mathbf{HOM}\,_G(\underline{V_i}, \xi) \right) \longrightarrow \xi$$

defined on each summand by formula

$$\alpha(x, \varphi) = \varphi(x), \ x \in V_i. \tag{3.24}$$

The mapping (3.24) is equivariant. To prove that (3.24) is an isomorphism it is sufficient to check it on each fiber. Here it follows from Schur's lemma. ∎

3.6 RELATIONS BETWEEN COMPLEX, SYMPLECTIC AND REAL BUNDLES

The category of G– equivariant vector bundles is good place to give consistent descriptions of three different structures on vector bundles — complex, real and symplectic.

Consider the group $G = \mathbf{Z}_2$ and a complex vector bundle ξ over the G–space X. This means that the group G acts on the space X. Let E be the total space of the bundle ξ and let

$$p : E \longrightarrow X$$

be the projection in the definition of the vector bundle ξ. Let G act on the total space E as a fiberwise operator which is linear over the reals and anticomplex over complex numbers, that is, if $\tau \in G = \mathbf{Z}_2$ is the generator then

$$\tau(\lambda x) = \bar{\lambda}\tau(x), \ \lambda \in \mathbf{C}, \ x \in E. \tag{3.25}$$

A vector bundle ξ with the action of the group G satisfying the condition (3.25) is called a KR–*bundle* . The operator τ is called the anticomplex involution. The corresponding Grothendieck group of KR–bundles is denoted by $KR(X)$.

Below we describe some of the relations with classical real and complex vector bundles.

Proposition 10 *Suppose that the G–space X has the form $X = Y \times \mathbf{Z}_2$ and the involution τ transposes the second factor. Then the group $KR(X)$ is naturally isomorphic to the group $K_U(Y)$ and this isomorphism coincides with restriction of a vector bundle to the summand $Y \times \{1\}$, $1 \in G = \mathbf{Z}_2$, ignoring the involution τ.*

Proof.

The complex bundle ξ over X consists of a union of two bundles: ξ_1 — restriction of ξ on $Y \times \{1\}$ and ξ_τ — restriction of ξ on $Y \times \{\tau\}$. Since τ interchanges the two summands — $Y \times \{1\}$ and $Y \times \{\tau\}$, the involution τ gives an isomorphism

$$\tau : \xi_1 \longrightarrow \xi_\tau$$

which is anticomplex. Hence

$$\xi_\tau = \tau^*(\overline{\xi_1}).$$

Hence given a complex vector bundle ξ_1 over $Y \times \{1\}$, then we can take the same vector bundle over $Y \times \{\tau\}$ with the conjugate complex structure. Then the identity involution τ is anticomplex and induces a KR– bundle over X. ∎

Proposition 11 *Suppose the involution τ on X is trivial. Then*

$$KR(X) \approx K_O(X). \tag{3.26}$$

The isomorphism (3.26) associates to any KR-bundle the fixed points of the involution τ.

Proposition 12 *The operation of forgetting the involution induces a homomorphism*

$$KR(X) \longrightarrow K_U(X)$$

and when the involution is trivial on the base X this homomorphism coincides with complexification

$$c : K_O(X) \longrightarrow K_U(X).$$

Similar to the realification of vector bundles, we can construct, for any base X with involution τ, a homomorphism

$$K_U(X) \longrightarrow KR(X) \tag{3.27}$$

which coincides with realification

$$r : K_U(X) \longrightarrow K_O(X),$$

when the involution on the base X is trivial. The homomorphism (3.27) associates to any complex vector bundle ξ over X the bundle $\eta = \xi \oplus \tau^*(\bar{\xi})$ with an involution which transposes the summands.

As for complex vector bundles the operation of tensor product can be defined for KR-bundles making $KR(X)$ into a ring. The extension of the groups $KR(X)$ to a cohomology theory is more interesting. An equivariant analogue of suspension for $\mathbf{Z}_2$-spaces must be defined.

Definition 8 *Denote by $D^{p,q}$ the disc D^{p+q} with involution τ which changes signs in the first p coordinates and denote by $S^{p,q}$ the boundary of $D^{p,q}$. Then the suspension of pair (X,Y) with involution is given by the pair*

$$\Sigma^{p,q}(X,Y) = (X \times D^{p,q}, (X \times S^{p,q}) \cup (Y \times D^{p,q}))$$

Hence in a natural way we can define a bigraded KR-theory

$$KR^{-p,-q}(X,Y) = KR\left(\Sigma^{p,q}(X,Y)\right), \tag{3.28}$$

where $KR(X,Y)$ is defined as the kernel of the mapping

$$KR(X/Y) \longrightarrow KR(pt).$$

Clearly the bigraded functor (3.28) is invariant with respect to homotopy and the following axiom holds: for any triple (X, Y, Z) we have the exact sequence

$$\ldots \longrightarrow KR^{p,q}(X, Y) \longrightarrow KR^{p,q}(X, Z) \longrightarrow KR^{p,q}(Y, Z) \longrightarrow \ldots$$
$$\ldots \longrightarrow KR^{p,q+1}(X, Y) \longrightarrow KR^{p,q+1}(X, Z) \longrightarrow KR^{p,q+1}(Y, Z) \longrightarrow = ts$$
$$\ldots \longrightarrow KR^{p,0}(X, Y) \longrightarrow KR^{p,0}(X, Z) \longrightarrow KR^{p,0}(Y, Z)$$

for any $p \leq 0$. As in classical K-theory, the operation of tensor product extends to a graded multiplication

$$KR^{p,q}(X, Y) \otimes KR^{p',q'}(X, Y') \longrightarrow KR^{p+p',q+q'}(X, Y \cup Y').$$

Moreover, the proof of Bott periodicity can be extended word by word to KR-theory:

Theorem 33 *There is an element $\beta \in KR(D^{1,1}, S^{1,1}) = KR^{-1,-1}(pt)$ such that the homomorphism given by multiplication by β*

$$\beta : KR^{p,q}(X, Y) \longrightarrow KR^{p-1,q-1}(X.Y) \tag{3.29}$$

is an isomorphism.

The proof is completely similar to that of Theorem 27 with the observation that all constructions are equivariant.

Further, the following facts are essential to our considerations.

Lemma 9 *Let X be space with involution. Then there exist equivariant homeomorphisms*

$$1)\, X \times S^{1,0} \times D^{0,1} \approx X \times S^{1,0} \times D^{1,0},$$
$$2)\, X \times S^{2,0} \times D^{0,2} \approx X \times S^{2,0} \times D^{2,0},$$
$$3)\, X \times S^{4,0} \times D^{0,4} \approx X \times S^{4,0} \times D^{4,0}.$$

In all three cases, the homeomorphism

$$\mu : X \times S^{p,0} \times D^{0,p} \longrightarrow X \times S^{p,0} \times D^{p,0}$$

is defined by the formula

$$\mu(x, s, u) = (x, s, su), \quad x \in X, s \in S^{p,0}, u \in D^{0,p},$$

where we identify $S^{p,0}$, $D^{p,0}$ and $D^{0,p}$, respectively, with the unit sphere or unit disk in the field of real, complex or quaternionic numbers, respectively.

Thus we have the following isomorphisms:

$$KR^{p,q}(X \times S^{1,0}) \approx KR^{p-1,q+1}(X \times S^{1,0}), \tag{3.30}$$

$$KR^{p,q}(X \times S^{2,0}) \approx KR^{p-2,q+2}(X \times S^{2,0}), \tag{3.31}$$

$$KR^{p,q}(X \times S^{4,0}) \approx KR^{p-4,q+4}(X \times S^{4,0}). \tag{3.32}$$

In the case of (3.30) using the Bott periodicity (3.29)

$$KR^{p,q}(X \times S^{1,0}) \approx KR^{p,q+2}(X \times S^{1,0}),$$

that is, the classical Bott periodicity for complex vector bundles.

The next step consists of the following theorem.

Theorem 34 *There is an element* $\alpha \in KR^{0,-8}(pt) = KR(D^{0,8}, S^{0,8})$ *such that the homomorphism given by multiplication by* α

$$\alpha : KR^{p,q}(X,Y) \longrightarrow KR^{p,q-8}(X,Y)$$

is an isomorphism.

In the case of the trivial action of an involution on the base we get the classical 8–periodicity in the real K–theory.

It turns out that this scheme can be modified so that it includes another type of K–theory – that of quaternionic vector bundles.

Let K be the(noncommutative) field of quaternions. As for real or complex vector bundles we can consider locally trivial vector bundles with fiber K^n and structure group $\mathbf{GL}(n,K)$, the so called quaternionic vector bundles. Each quaternionic vector bundle can be considered as a complex vector bundle $p : E \longrightarrow X$ with additional structure defined by a fiberwise anticomplex linear operator J such that

$$J^2 = -1, IJ + JI = 0,$$

where I is fiberwise multiplication by the imaginary unit.

More generally, let J be a fiberwise anticomplex linear operator which acts on a complex vector bundles ξ and satisfies

$$J^4 = 1, IJ + JI = 0. \tag{3.33}$$

Then the vector bundle ξ can be split into two summands $\xi = \xi_1 \oplus \xi_2$ both invariant under action of J, that is, $J = J_1 \oplus J_2$ such that

$$J_1^2 = 1, \quad J_2^2 = -1. \tag{3.34}$$

Hence the vector bundle ξ_1 is the complexification of a real vector bundle and ξ_2 is quaternionic vector bundle.

Consider a similar situation over a base X with an involution τ such that the operator (3.33) commutes with τ. Such vector bundle will be called a *KRS-bundle* .

Lemma 10 *A KRS–bundle ξ is split into an equivariant direct sum $\xi = \xi_1 \oplus \xi_2$ such that $J^2 = 1$ on ξ_1 and $J^2 = -1$ on ξ_2.*

Lemma 10 shows that the Grothendieck group $KRS(X)$ generated by $KRS-$ bundles has a $\mathbf{Z}_2$–grading, that is,

$$KRS(X) = KRS_0(X) \oplus KRS_1(X).$$

It is clear that $KRS_0(X) = KR(X)$. In the case when the involution τ acts trivially, $KRS_1(X) = K_Q(X)$, that is,

$$KRS(X) = K_O(X) \oplus K_Q(X)$$

where $K_Q(X)$ is the group generated by quaternionic bundles.

The $\mathbf{Z}_2$–graded group $KRS(X)$ has a multiplicative structure induced by the . tensor product of complex vector bundles. It is clear that the tensor product preserves the action of involution τ on the base and action of the operator J on the total space. Moreover, this action is compatible with the grading, that is, if $\xi_1 \in KRS_i(X)$, $\xi_2 \in KRS_j(X)$ then $\xi_1 \otimes \xi_2 \in KRS_{i+j}(X)$. As in KR–theory, the groups $KRS^{p,q}(X,Y) = KRS_0^{p,q}(X,Y) \oplus KRS_1^{p,q}(X,Y)$ may be defined:

$$KRS^{-p,-q}(X,Y) = KRS\left(D^{p,q} \times X, (S^{p,q} \times X) \cup (D^{p,q} \times Y)\right),$$

where

$$KRS(X,Y) = \mathbf{Ker}\ (KRS(X/Y) {\longrightarrow} KRS(\mathrm{pt}))\,.$$

Bott periodicity generalizes: if $\beta \in KR\left(D^{1,1}, S^{1,1}\right) = KRS_0^{-1,-1}(\mathrm{pt})$ is the Bott element then

Theorem 35 *The homomorphism*

$$\beta : KRS^{p,q}(X,Y) \longrightarrow KRS^{p-1,q-1}(X,Y)$$

given by multiplication by the element β is an isomorphism.

As in Theorem 34 we have the following

Theorem 36 *There is an element*

$$\gamma \in KRS_1^{0,-4}(pt) = KRS_1(D^{0,4}, S^{0,4})$$

such that the homomorphism of given by multiplication by γ

$$\gamma : KRS^{p,q}(X,Y) \longrightarrow KRS^{p,q-4}(X,Y)$$

or

$$\gamma : KRS_0^{p,q}(X,Y) \longrightarrow KRS_1^{p,q-4}(X,Y)$$

or

$$\gamma : KRS_1^{p,q}(X,Y) \longrightarrow KRS_0^{p,q-4}(X,Y)$$

are isomorphisms. Moreover,

$$\gamma^2 = \alpha \in KRS^{0,-8}(pt) = KR^{0,-8}(pt) = K_O^{-8}(pt).$$

Theorem 36 says that $\gamma \in K_Q^{-4}(\mathrm{pt})$. In particular, we have the following isomorphisms:

$$K_O^q(X) \approx K_Q^{q-4}(X)$$
$$K_Q^q(X) \approx K_O^{q-4}(X).$$

q	-8	-7	-6	-5	-4	-3	-2	-1	0
K_O	**Z** $\alpha = \gamma^2$	0	0	0	**Z** $u\gamma$	0	$\mathbf{Z}_2$ h^2	$\mathbf{Z}_2$ h	**Z** $u^2 = 4$
K_Q	**Z** $u\gamma^2$	0	$\mathbf{Z}_2$ $h^2\gamma$	$\mathbf{Z}_2$ $h\gamma$	**Z** γ	0	0	0	**Z** u

Figure 3.1

A list of the groups K_O and K_Q are given in the figure 3.1.

4

CALCULATION METHODS IN
K–THEORY

In this chapter we shall give some methods for describing the K–groups of various concrete spaces by reducing them to a description in terms of the usual cohomology groups of spaces. The methods we discuss are spectral sequences, cohomology operations and direct images. These methods cannot be used universally but they allow us to use certain geometric properties of our concrete topological spaces and manifolds.

4.1 SPECTRAL SEQUENCES

Similar to classical cohomology theory, the spectral sequence for K–theory is constructed using a filtration of a space X. The construction of the spectral sequence described below can be applied not only to K–theory but to any generalized cohomology theory. Thus let

$$X_0 \subset X_1 \subset \ldots \subset X_N = X \qquad (4.1)$$

be a increasing filtration of the space X. Put

$$D = \bigoplus_{p,q} D^{p,q} = \bigoplus_{p,q} K^{p+q}(X_p)$$

$$E = \bigoplus_{p,q} E^{p,q} = \bigoplus_{p,q} K^{p+q}(X_p, X_{p-1}). \qquad (4.2)$$

Consider the exact sequence of a pair (X_p, X_{p-1}):

$$\ldots \longrightarrow K^{p-1+q}(X_{p-1}) \overset{\partial}{\longrightarrow} K^{p+q}(X_p, X_{p-1}) \overset{j^*}{\longrightarrow}$$
$$\longrightarrow K^{p+q}(X_p) \overset{i^*}{\longrightarrow} K^{p+q}(X_{p-1}) \longrightarrow \ldots \qquad (4.3)$$

After summation by all p and q, the sequences

$$\ldots \longrightarrow \bigoplus D^{p-1,q} \xrightarrow{\partial} \bigoplus E^{p,q} \xrightarrow{j^*} \bigoplus D^{p,q} \xrightarrow{i^*} \bigoplus D^{p-1,q+1} \longrightarrow \ldots ,$$

can be written briefly as

$$\ldots \longrightarrow D \xrightarrow{\partial} E \xrightarrow{j^*} D \xrightarrow{i^*} D \xrightarrow{\partial} E \longrightarrow \ldots , \tag{4.4}$$

where the bigradings of the homomorphisms ∂, j^*, i^* are as follows:

$$\begin{aligned}
\deg i^* &= (-1,1) \\
\deg j^* &= (0,0) \\
\deg \partial &= (1,0).
\end{aligned}$$

The sequence (4.4) can be written as an exact triangle

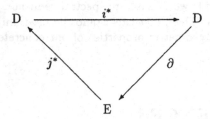

$$\tag{4.5}$$

Put

$$d = \partial j^* : E \longrightarrow E.$$

Then the bigrading of d is $\deg d = (1,0)$ and it is clear that $d^2 = 0$. Put

$$\begin{aligned}
D_1 &= D, \\
E_1 &= E, \\
i_1 &= i^*, \\
j_1 &= j^*, \\
\partial_1 &= \partial, \\
d_1 &= d.
\end{aligned}$$

Now put

$$\begin{aligned}
D_2 &= i_1(D_1) \subset D_1, \\
E_2 &= H(E_1, d_1).
\end{aligned}$$

The grading of D_2 is inherited from the grading as an image, that is,

$$D_2 = \bigoplus D_2^{p,q},$$
$$D_2^{p,q} = i_1(D_1^{p+1,q-1}).$$

The grading of E_2 is inherited from E_1. Then we put

$$i_2 = i_1|_{D_2} : D_2 \longrightarrow D_2,$$
$$\partial_2 = \partial_1 i_1^{-1},$$
$$j_2 = j_1.$$

It is clear that i_2, j_2 and ∂_2 are well defined. In fact, if $x \in D_2$ then $i_1(x) \in D_2$. If $x \in D_1$ then $x = i_1(y)$ and

$$\partial_2(x) = [\partial_1(y)] \in H(E_1, d_1).$$

The latter inclusion follows from the identity

$$d_1 \partial_1(y) = \partial_1 j_1 \partial_1(y) = 0.$$

If $i_1(y) = 0$, the exactness of the triangle (4.5) gives $y = j_1(x)$ and then .

$$\partial_2(x) = [\partial_1 j_1(z)] = [d_1(z)] = 0.$$

Hence ∂_2 is well defined. Finally, if $x \in E_1$, $d_1(x) = 0$ then $\partial_1 j_1(x) = 0$. Hence $j_1(x) = i_1(z) \in D_2$. If $x = d_1(y)$, $x = \partial_1 j_1(z)$ then $j_1(x) = j_1 \partial_1 j_1(z) = 0$ Hence j_2 is well defined. Thus there is a new exact triangle

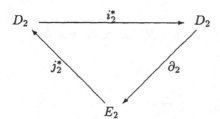

This triangle is said to be derived from the triangle (4.5).

Repeating the process one can construct a series of exact triangles

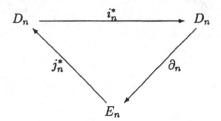

The bigradings of homomorphisms i_n, j_n, ∂_n and $d_n = \partial_n j_n$ are as follows:

$$\begin{aligned}
\deg i_n &= (-1, 1), \\
\deg j_n &= (0, 0), \\
\deg \partial_n &= (n, -n + 1), \\
\deg d_n &= (n, -n + 1).
\end{aligned}$$

The sequence

$$(E_n, d_n) \tag{4.6}$$

is called the *spectral sequence in K-theory* associated with a filtration (4.1).

Theorem 37 *The spectral sequence (4.6) converges to the groups associated to the group $K^*(X)$ by the filtration (4.1).*

Proof.

According to the definition (4.2), $E_1^{p,q} = 0$ for $p > N$. Hence for $n > N$,

$$d_n = 0,$$

that is,

$$E_n^{p,q} = E_{n+1}^{p,q} = \ldots = E_\infty^{p,q}.$$

By definition, we have

$$D_n^{p,q} = \mathbf{Im}\left(K^{p+q}(X_{p+n}) \longrightarrow K^{p+q}(X_p)\right).$$

Hence for $n > N$,

$$D_n^{p,q} = \text{Im} \left(K^{p+q}(X) \longrightarrow K^{p+q}(X_p) \right).$$

Hence the homomorphism i_n is an epimorphism. Thus we have the exact sequence

$$0 \longrightarrow E_n^{p,q} \xrightarrow{j_n} D_n^{p,q} \xrightarrow{i_n} D_n^{p-1,q+1} \longrightarrow 0.$$

Hence

$$E_n^{p,q} = \text{Ker } i_n =$$
$$= \text{Ker} \left(K^{p+q}(X) \longrightarrow K^{p+q}(X_{p-1}) \right) / \text{Ker} \left(K^{p+q}(X) \longrightarrow K^{p+q}(X_p) \right)$$

∎

Let $p : Y \longrightarrow X$ be a locally trivial bundle with fibre F. Then the fundamental group $\pi_1(X, x_0)$ acts on the fiber F in the sense that there is natural homomorphism

$$\pi_1(X, x_0) \longrightarrow [F, F], \tag{4.7}$$

where $[F, F]$ is the family of homotopy equivalences of F. Then the homomorphism (4.7) induces an action of the group $\pi_1(X, x_0)$ on the groups $K^*(F)$.

Theorem 38 *Let $p : Y \longrightarrow X$ be a locally trivial bundle with fiber F and the trivial action of $\pi_1(X, x_0)$ on K-groups of the fiber. Then the spectral sequence generated by filtration $Y_k = p^{-1}([X]^k)$, where $[X]^k$ is k-dimensional skeleton of X, converges to the groups associated to $K^*(Y)$ and the second term has the following form:*

$$E_2^{p,q} = H^p \left(X, K^q(F) \right).$$

Corollary 6 *If the filtration of X is formed by the k-dimensional skeletons then the corresponding spectral sequence converges to the groups associated to $K^*(X)$ and*

$$E_2^{p,q} = \begin{cases} H^p(X; Z), & \text{for even } q \\ 0 & \text{for odd } q. \end{cases} \tag{4.8}$$

Proof.

The first term of the spectral sequence E_1 is defined to be

$$E_1^{p,q} = K^{p+q}(Y_p, Y_{p-1}).$$

Since the locally trivial bundle is trivial over each cell, the pair (Y_p, Y_{p-1}) has the same K–groups as the union

$$\bigcup_j (\sigma_j^p \times F, \partial\sigma_j^p \times F),$$

that is,

$$E_1^{p,q} \approx \bigoplus_j K^{p,q}(\sigma_j^p \times F, \partial\sigma_j^p \times F) = \bigoplus_j K^q(F).$$

Hence, we can identify the term E_1 with the cochain group

$$E_1^{p,q} = C^p(X; K^q(F)) \tag{4.9}$$

with coefficients in the group $K_q(F)$. What we need to establish is that the differential d_1 coincides with the coboundary homomorphism in the chain groups of the space X. This coincidence follows from the exact sequence (4.3). Notice that the coincidence (4.9) only holds if the fundamental group of the base X acts trivially in the K–groups of the fiber F. In general, the term E_1 is isomorphic to the chain group with a local system of coefficients defined by the action of the fundamental group $\pi_1(X, x_0)$ in the group $K^q(F)$.

We say that the spectral sequence is *multiplicative* if all groups $E_s = \bigoplus_{p,q} E_s^{p,q}$ are bigraded rings, the differentials d_s are derivations, that is,

$$d_s(xy) = (d_s x)y + (-1)^{p+q} x(d_s y), \; x \in E_s^{p+q},$$

and the homology of d_s is isomorphic to E_{s+1} as a ring.

Theorem 39 *The spectral sequence from Theorem 38 is multiplicative. The isomorphism (4.8) is an isomorphism of rings. The ring structure of E_∞ is isomorphic to the ring structure of groups associated with the filtration generated by the skeletons of the base X.*

We leave the proof of this theorem to the reader.

Examples

1. Let $X = \mathbf{S}^n$. Then the E_2 term of the spectral sequence is

$$E_2^{p,q} = H^p(\mathbf{S}^n; K^q(*)) = \begin{cases} \mathbf{Z} & \text{for } p = 0, n \text{ and for even } q, \\ 0 & \text{in other cases.} \end{cases} \left.\right\} \text{ (see fig.4.1.)}$$

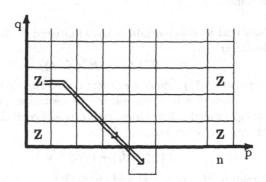

Figure 4.1

For a one point space all differentials are trivial. Hence

$$d_s|_{E_s^{o,q}} = 0.$$

Thus all differentials for a sphere are trivial and $E_2 = E_\infty$. Finally, we have for even n,

$$K^0(\mathbf{S}^n) = \mathbf{Z} \oplus \mathbf{Z}, \ K^1(\mathbf{S}^n) - 0,$$

and for odd n,

$$K^0(\mathbf{S}^n) = \mathbf{Z}, \ K^1(\mathbf{S}^n) = \mathbf{Z}.$$

2. Let $X = \mathbf{CP}^n$. Then

$$E_2^{**} = H^*(\mathbf{CP}^n; K^*(\mathrm{pt})) = \mathbf{Z}[t]/\{t^{n+1} = 0\}.$$

Nontrivial terms $E_2^{p,q}$ exist only when $p + q$ is even. Hence all differentials are trivial and we have

$$E_\infty^{**} = \mathbf{Z}[t]/\{t^{n+1} = 0\}. \tag{4.10}$$

Hence the additive structure of the group $K(\mathbf{CP}^n)$ is the same as in the term E_∞. To show that the multiplicative structure of the group $K^*(\mathbf{CP}^n)$ coincides with (4.10), consider an element $u \in K(\mathbf{CP}^n)$ which corresponds to the element $t \in E_\infty^{2,0}$. The element u should be trivial on the one dimensional skeleton of $\mathbf{CP}^n$. Further, the element u^k, $k \leq n$ is trivial on the $(2k - 1)$–dimensional skeleton of $\mathbf{CP}^n$ and corresponds to the element $t^k \in E_\infty^{2k,0}$. Consider the element $u^{n+1} \in K(\mathbf{CP}^n)$. This element is trivial on the $(2n + 1)$–dimensional skeleton, that is, $u^{n+1} = 0$. Hence

$$K^*(\mathbf{CP}^n) = \mathbf{Z}[u]/\{u^{n+1} = 0\}.$$

It is easy to show that the element u can be chosen to be equal to $[\xi] - 1$ where ξ is the Hopf bundle.

3. Let us show that all differentials d_s have finite order. To prove this consider the Chern character

$$\mathrm{ch} \; : K^*(X) \longrightarrow H^*(X; \; \mathbf{Q}). \qquad (4.11)$$

The homomorphism (4.11) defines a natural transformation of cohomology theories and hence induces a homomorphism of spectral sequences generated by K–theory and $\mathbf{Z}_2$–graded cohomology with rational coefficients induced by the filtration with respect to skeletons of X:

$$\mathrm{ch} \; : (E_s^{p,q}, d_s) \longrightarrow ('E_s^{p,q}, 'd_s).$$

Since all differentials $'d_s$ are trivial and $\mathrm{ch} \otimes QQ$ is an isomorphism, we have

$$d_s \otimes \mathbf{Q} = 0.$$

Thus the differential d_s has finite order on each group $E_s^{p,q}$.

For example, the differential

$$d_2 : E_2^{p,q} \longrightarrow E_s^{p+2,q-1}$$

can be written as

$$d_2 : H^p\left(X; \; K^q(*)\right) \longrightarrow H^{n+2}\left(X; \; K^{k-1}(*)\right).$$

Hence $d_2 = 0$ and $E_2^{p,q} = E_3^{p,q}$. Thus the differential d_3 has the form of a stable cohomology operation:

$$d_3 : H^p(X; \; \mathbf{Z}) \longrightarrow H^{p+3}(X; \; \mathbf{Z}).$$

Therefore, d_3 may take one of only two possible values: $d_3 = 0$ or $d_3 = \mathrm{Sq}^3$. To check which possibility is realized it suffices to consider a special example for the space X where the differential d_3 can be calculated.

4. As in cohomology theory we can obtain a formula for the K–groups of a Cartesian product $X \times Y$. Assume firstly that the group $K(Y)$ has no torsion. Let $X_1 \subset X$ be a subspace of X which differs from X in only one cell. Consider the exact sequences for pairs (X, X_1) and $(X \times Y, X_1 \times Y)$:

$$\cdots \longrightarrow K^*(X, X_1) \otimes K^*(Y) \longrightarrow K^*(X) \otimes K^*(Y) \longrightarrow K^*(X_1) \otimes K^*(Y) \longrightarrow \cdots$$

$$\Big\downarrow f \qquad\qquad\qquad \Big\downarrow g \qquad\qquad\qquad \Big\downarrow h$$

$$\cdots \longrightarrow K^*(X \times Y, X_1 \times Y) \longrightarrow K^*(X \times Y) \longrightarrow K^*(X_1 \times Y) \longrightarrow \cdots \qquad (4.12)$$

The upper row is exact since $K^*(Y)$ has no torsion. The vertical homomorphisms are defined by tensor multiplication. By Bott periodicity, the vertical homomorphism f is an isomorphism. Hence using induction and the Five Lemma for isomorphisms we get that g is an isomorphism.

Now let Y_1 be a space such that $K^*(Y_1)$ has no torsion and suppose there is a map

$$f : Y \longrightarrow Y_1$$

such that the homomorphism

$$f^* : K^*(Y_1) \longrightarrow K^*(Y)$$

is an epimorphism. Then the exact sequence for the pair (Y_1, Y) has the form of a free resolution

$$0 \longrightarrow K^*(Y_1, Y) \longrightarrow K^*(Y_1) \longrightarrow K^*(Y) \longrightarrow 0.$$

Consider the diagram

$$\cdots \longrightarrow K^*(Y_1, Y) \otimes K^*(X) \overset{\beta}{\longrightarrow} K^*(Y_1) \otimes K^*(X) \overset{\alpha}{\longrightarrow} K^*(Y) \otimes K^*(X) \longrightarrow 0$$
$$\downarrow{f} \qquad\qquad\qquad \downarrow{g} \qquad\qquad\qquad \downarrow{h}$$
$$\cdots \longrightarrow K^*(Y_1 \times X, Y \times X) \overset{\beta'}{\longrightarrow} K^*(Y_1 \times X) \overset{\alpha'}{\longrightarrow} K^*(Y \times X) \longrightarrow \cdots$$

We will show that the homomorphism f is a monomorphism. If $f(x) = 0$ then there exists y such that $\alpha(y) = x$. Then $\alpha'(g(y)) = 0$. So there exists z such that $g(y) = \beta'(z)$. Since h is an isomorphism, we have $z = h(w)$ or $g(y) = g\beta(w)$. Since g is also an isomorphism, $y = \beta(w)$. Hence $x = \alpha(w) = 0$.

Now let us show that $h(\mathbf{Ker}\ \beta) = \mathbf{Ker}\ \beta'$. It is clear that $h(\mathbf{Ker}\ \beta) \subset \mathbf{Ker}\ \beta'$. If $\beta'(x) = 0$, $x = h(y)$ then $g\beta(y) = \beta'h(y) = 0$, that is, $\beta(y) = 0$ or $y \in \mathbf{Ker}\ \beta$. Hence there is an exact sequence

$$0 \longrightarrow K^*(Y) \otimes K^*(X) \longrightarrow K^*(Y \times X) \longrightarrow \mathbf{Ker}\ \beta \longrightarrow 0.$$

The group $\mathbf{Ker}\ \beta$ is denoted by $\mathbf{tor}\ (K^*(Y), K^*(X))$. So there is a Künneth formula

$$0 \longrightarrow K^*(Y) \otimes K^*(X) \longrightarrow K^*(Y \times X) \longrightarrow \mathbf{Tor}\ (K^*(Y), K^*(X)) \longrightarrow 0.$$

To complete the proof it remains for us to construct the space Y_1. The space Y_1 one should regard as a Cartesian product of complex Grassmannian manifolds which classify finite families of generators of the group $K^*(Y)$. The cohomology of the Grassmannian manifolds has no torsion. Thus the proof is complete. ∎

5. *The splitting principle. If a functorial relation holds for vector bundles which split into a sum of one dimensional bundles then the same relation holds for all vector bundles.*

This principle is based on the following theorem

Theorem 40 *For any vector bundle ξ over the base X there is a continuous mapping $f : Y \longrightarrow X$ such that $f^*(\xi)$ splits into a sum of one dimensional bundles and the homomorphism*

$$f^* : K^*(X) \longrightarrow K^*(Y)$$

is a monomorphism.

Proof.

Let $p : E \longrightarrow X$ give the vector bundle ξ. Denote by $P(E)$ the space of all one dimensional subspaces of fibers of the ξ. Let $\pi : P(E) \longrightarrow X$ be the natural mapping. Then the inverse image $\pi^*(\xi)$ split into sum $\pi^*(\xi) = \eta \oplus \zeta$ where η is the Hopf bundle over $P(E)$. Hence $\dim \zeta = \dim \xi - 1$. On the other hand, the homomorphism

$$\pi^* : K^*(X) \longrightarrow K^*(P(E))$$

is a monomorphism. In fact, consider the continuous mapping $f : X \longrightarrow \mathbf{BU}(n)$ and commutative diagram

$$
\begin{array}{ccc}
P(E) & \longrightarrow & PEU(n) \\
\downarrow & & \downarrow \\
X & \longrightarrow & \mathbf{BU}(n)
\end{array}
\qquad (4.13)
$$

The diagram (4.13) induces a commutative diagram of spectral sequences

where

$$
\begin{aligned}
{}^{1}E_{s}^{p,q} &= H^{p}\left(X, K^{q}(\mathbf{CP}^{n-1})\right), \\
{}^{2}E_{s}^{p,q} &= H^{p}\left(\mathbf{BU}(n), K^{q}(\mathbf{CP}^{n-1})\right), \\
{}^{3}E_{s}^{p,q} &= H^{p}\left(X, K^{q}(\mathrm{pt})\right), \\
{}^{4}E_{s}^{p,q} &= H^{p}\left(\mathbf{BU}(n), K^{q}(\mathrm{pt})\right).
\end{aligned}
$$

In the spectral sequences ${}^{2}E$ and ${}^{4}E$, all differentials are trivial. Hence the differentials in ${}^{1}E$ are induced by the differentials in ${}^{3}E$. Thus

$$
{}^{1}E_{s}^{*,*} = {}^{3}E_{s}^{*,*} \otimes K^{*}(\mathbf{CP}^{n-1}).
$$

Hence the homomorphism

$$
{}^{3}E_{\infty}^{*,*} \longrightarrow {}^{1}E_{\infty}^{*,*}
$$

is a monomorphism. Therefore, π^{*} is also a monomorphism. Now one can complete the proof by induction with respect to the dimension of vector bundle ξ. ∎

4.2   OPERATIONS IN K–THEORY

By definition, a cohomology operation is a transformation

$$
\alpha : K^{*}(X) \longrightarrow K^{*}(X)
$$

which is functorial, that is, for any continuous mapping $f : X \longrightarrow Y$ the following diagram

$$
\begin{array}{ccc}
K^*(X) & \xleftarrow{\;f^*\;} & K^*(Y) \\
\Big\downarrow{\alpha} & & \Big\downarrow{\alpha} \\
K^*(X) & \xleftarrow{\;f^*\;} & K^*(Y)
\end{array}
$$

is commutative. In particular, if $f : X \longrightarrow \mathbf{BU}(N)$, $x = f^*(\xi_N)$ then

$$\alpha(x) = f^*(\alpha(\xi_N)).$$

Here the element $\alpha(\xi_N) \in K^*(\mathbf{BU}(N))$ may be an arbitrary element and it completely defines the cohomology operation α. This definition is not quite correct because the classifying space $\mathbf{BU}(N)$ is not a finite cellular complex. Strictly speaking, we should consider the inverse sequence of the Grassmann manifolds $G(N + n, N)$, $n \longrightarrow \infty$ with natural inclusions

$$j_n : G(N + n, N) \longrightarrow G(N + n + 1, N).$$

Then the cohomology operation α is defined as an inverse sequence of elements $\alpha_n \in K^*(G(N+n, N))$ such that $j_n^*(\alpha_{n+1}) = \alpha_n$. This means that the sequence $\{\alpha_n\}$ defines an element in the inverse limit

$$\alpha \in \varprojlim_{n} K^*(G(N + n, N)) = \mathcal{K}^*(\mathbf{BU}(N)). \tag{4.14}$$

The expression (4.14) can be considered as a definition of the K–groups of an infinite cellular spaces.

Let us continue the analysis of cohomology operations. Each cohomology operation should be independent of trivial direct summands. This means that a cohomology operation is a sequence of elements $\alpha_N \in \mathcal{K}^*(\mathbf{BU}(N))$ such that

$$\alpha_{N+1}(\xi \oplus 1) = \alpha_N(\xi).$$

Then the natural mapping

$$k_N : \mathbf{BU}(N) \longrightarrow \mathbf{BU}(N + 1)$$

sends the element α_{N+1} to α_N. This means that the sequence $\{\alpha_N\}$ defines an element

$$\alpha \in \varprojlim_{N} \mathcal{K}^*(\mathbf{BU}(N)).$$

Let us put $\mathbf{BU} = \varinjlim_{N} \mathbf{BU}(N)$. Then we can show that

$$\mathcal{K}^*(\mathbf{BU}) = \varprojlim_{N} \mathcal{K}^*(\mathbf{BU}(N)).$$

Definition 9 *The Adams operation* $\Psi^k : K(X) \longrightarrow K(X)$ *is the operation characterized by the following property* .

$$\Psi^k(\xi_1 \oplus \ldots \xi_n) = \xi_1^k \oplus \ldots \xi_n^k$$

for any one dimensional vector bundles ξ_j.

We show now that such an operation Ψ^k exists. Consider the symmetric polynomial $s_k(t_1, \ldots, t_n) = t_1^k + \ldots + t_n^k$. Then the polynomial s_k can be expressed as a polynomial

$$s_k(t_1, \ldots, t_n) = P_k^n(\sigma_1, \ldots, \sigma_n).$$

The polynomial P_k^n has integer coefficients and the σ_k are the elementary symmetric polynomials:

$$P_k^n(\sigma_1, \ldots, \sigma_n) = \bigoplus \lambda_{l_1, l_2, \ldots, l_n}^k \sigma_1^{l_1} \sigma_2^{l_2} \cdots \sigma_n^{l_n}.$$

Then

$$\Psi^k(\xi) = \bigoplus \lambda_{l_1, l_2, \ldots, l_n}^k \xi^{l_1} \otimes (\Lambda_2 \xi)^{l_2} \otimes \ldots \otimes (\Lambda_n \xi)^{l_n}$$

It is easy to check that

$$\Psi^k(\xi_1 \oplus \xi_2) = \Psi^k(\xi_1) + \Psi^k(\xi_2).$$

Theorem 41 *The Adams operations have the following properties:*

1. $\Psi^k(xy) = \Psi^k(x)\Psi^k(y)$;

2. $\Psi^k \Psi^l = \Psi^{kl}$;

3. $\Psi^p(x) \equiv x^p (\mod p)$ *for prime p.*

Proof.

As was shown earlier, it is sufficient to check the identities on vector bundles which can be split into one–dimensional summands. Let $\xi = \xi_1 \oplus \ldots \oplus \xi_n$, $\eta = \eta_1 \oplus \ldots \oplus \eta_m$ and then

$$\xi \otimes \eta \bigoplus_{i,j} (\xi_i \otimes \eta_j).$$

Then we have

$$\Psi^k(\xi \otimes \eta) = \bigoplus_{i,j} (\xi_i \otimes \eta_j)^k =$$

$$= \bigoplus_{i,j} (\xi_i^k \otimes \eta_j^k) =$$

$$= \left(\bigoplus_i \xi_i^k \right) \otimes \left(\bigoplus_j \eta_j^k \right) =$$

$$= \Psi^k(\xi) \Psi^k(\eta).$$

Further,

$$\Psi^k \Psi^l (\xi_1 \oplus \ldots \oplus \xi_n) = \Psi^k (\xi_1^l \oplus \ldots \oplus \xi_n^l) =$$
$$= \xi_1^{kl} \oplus \ldots \oplus \xi_n^{kl} = \Psi^{kl} (\xi_1 \oplus \ldots \oplus \xi_n).$$

Finally, if p is a prime integer then,

$$\Psi^p (\xi_1 \oplus \ldots \oplus \xi_n) = \xi_1^p \oplus \ldots \oplus \xi_n^p \equiv$$
$$\equiv (\xi_1 \oplus \ldots \oplus \xi_n)^p \ (\bmod\ p).$$

Examples

1.   Consider the sphere $\mathbf{S}^{2n}$ and the generator $\beta_n \in K^0(\mathbf{S}^{2n})$. The element β_n is the image of the element

$$\beta \otimes \ldots \otimes \beta \in K^0(\mathbf{S}^2 \times \ldots \times \mathbf{S}^2)$$

by the homomorphism induced by the continuous mapping

$$f : \mathbf{S}^2 \times \ldots \times \mathbf{S}^2 \longrightarrow \mathbf{S}^{2n}$$

which contracts the $(2n-2)$–dimensional skeleton to a point.

Then

$$f^* \Psi^k (\beta_n) \, \Psi^k (\beta \otimes \ldots \otimes \beta) \, (\Psi^k \beta) \otimes \ldots \otimes (\Psi^k \beta) \, .$$

Now $\beta = \eta - 1$ where η is the Hopf bundle. Hence

$$\Psi^k \beta = \eta^k - 1 = (1 + \beta)^k - 1 = k\beta.$$

Thus

$$f^* \Psi^k \beta_n = k^n (\beta \otimes \ldots \otimes \beta) \, k^n f^* (\beta_n),$$

that is,

$$\Psi^k \beta_n = k^n \beta_n \, .$$

2. Division algebras.

Let V be a finite dimensional division algebra, that is, any $x \in V$, $x \neq 0$, has a multiplicative inverse $x^{-1} \in V$. Denote by $P(V^3)$ the set of all subspaces of $V^3 = V \oplus V \oplus V$ which are one dimensional V–modules. The space $P(V^3)$ is called the (2–dimensional) projective space over the algebra V. We know of three cases of division algebras: the real and complex fields, and the skew field of quaternions.

We can calculate the cohomology of the space $P(V^3)$. Let $n = 2m = \dim V > 1$ be even. There is a locally trivial bundle

$$\mathbf{S}^{3n-1} \longrightarrow P(V^3)$$

with the fiber $\mathbf{S}^{n-1}$. As for $\mathbf{CP}^n$ (see 18), we can use the spectral sequence to show that

$$H^* \left(P(V^3) \right) = \mathbf{Z}[v]/\{v^3 = 0\}, \, \dim v = n.$$

Let us use the K–theory spectral sequence for the space $P(V^3)$. We have

$$K^* \left(P(V^3) \right) = \mathbf{Z}[u]/\{u^3 = 0\}, \, u \in K^0 \left(P(V^3) \right) .$$

Moreover, the space $P(V^3)$ is the union of the sphere $\mathbf{S}^n = P(V^2)$ and the disk $\mathbf{D}^{2n}$ glued on by a continuous mapping $\varphi : \partial \mathbf{D}^{2n} = \mathbf{S}^{2n-1} \longrightarrow \mathbf{S}^n$. In other words, we have the pair

$$\mathbf{S}^n \subset P(V^3) \longrightarrow P(V^3)/\mathbf{S}^n = \mathbf{S}^{2n}. \qquad (4.15)$$

The exact sequence for (4.15) has the following form

$$0 \longrightarrow K^0 \left(\mathbf{S}^{2n} \right) \xrightarrow{j^*} K^0 \left(P(V^3) \right) \xrightarrow{i^*} K^0 \left(\mathbf{S}^n \right) \longrightarrow 0.$$

Let

$$a \in K^0\left(\mathbf{S}^n\right); \; b \in K^0\left(\mathbf{S}^{2n}\right)$$

be generators and let

$$i^*\left(x\right) = a; \; j^*\left(b\right) = y.$$

Let $y = \alpha u + \beta u^2$. Then $0 = y^2 = \left(\alpha u + \beta u^2\right)^2 \alpha^2 u^2$. Hence $\alpha = 0$. Let $x = \gamma u + \delta u^2$. The elements (x, y) forms the basis in the group $K^0\left(P(V^3)\right)$ and so $\beta = \pm 1$; $\gamma = \pm 1$. By choosing a and b appropriately we can assume that

$$x^2 = y.$$

Now we can calculate the action of the Adams operations. Firstly,

$$j^* \Psi^k\left(x\right) = \Psi^k\left(a\right) = k^m a.$$

Hence

$$\Psi^k\left(x\right) = k^m x + \lambda y$$

and

$$\Psi^k\left(y\right) = \Psi^k j_*(b) = j^* \Psi^k\left(b\right) = k^{2m} y.$$

Moreover,

$$\Psi^k\left(x\right) \equiv x^k\left(\bmod k\right)$$

when k is prime. For example, when $k = 2$, the number λ must be odd; when $k = 3$, the number λ is divisible by 3.

Consider the operation $\Psi^6 = \Psi^2 \Psi^3 = \Psi^3 \Psi^2$. Then

$$\Psi^2 \Psi^3(x) = \Psi^2\left(3^m x + \mu y\right) 6^m x + \left(3^m \lambda + 2^{2m} \mu\right) y,$$
$$\Psi^3 \Psi^2(x) = \Psi^3\left(2^m x + \lambda y\right) 6^m x + \left(3^{2m} \lambda + 2^m \mu\right) y,$$

or

$$2^m\left(2^m - 1\right)\mu = 3^m\left(3^m - 1\right)\lambda,$$

where λ is odd. This means that the number $(3^m - 1)$ is divisible by 2^m. Let $m = 2^k l$ where l is odd. Then

$$3^m - 1 = \left(3^{2^k}\right)^l - 1\left(3^{2^k} - 1\right)\left(\left(3^{2^k}\right)^{l-1} + \left(3^{2^k}\right)^{l-2} + \ldots + 1\right)$$

The second factor has an odd number of summands and, therefore, it is odd. Hence the number $3^{2^k} - 1$ is divisible by 2^m and hence is divisible by 2^{2^k}. On the other hand,

$$3^{2^k} - 1 = \left(3^{2^{k-1}} + 1\right)\left(3^{2^{k-2}} + 1\right) \cdot \ldots \cdot (3 + 1)(3 - 1). \tag{4.16}$$

It is easy to see that the number $3^{2^k} + 1$ is divisible by 2 but not divisible by 4. Hence by the decomposition (4.16), the number $3^{2^k} - 1$ is divisible by 2^{k+2} but not divisible by 2^{k+3}. Hence

$$2^k l \leq k + 2. \tag{4.17}$$

For $l = 1$, the inequality (4.17) holds for $k = 0, 1, 2$. For $l \geq 3$, (4.17) has no solution. Hence division algebras exist only in dimensions $n = 2, 4, 8$. In fact, in addition to the real, complex and quaternionic fields, there is another 8–dimensional (nonassociative) algebra, the Cayley numbers. Even though the algebra of the Cayley numbers is not associative, the projective space $P(V^3)$ can be defined.

It remains to consider an odd dimensional division algebra. But in this case, we must have $\dim V = 1$ since on an even dimensional sphere there are no nonvanishing vector fields.

4.3   THE THOM ISOMORPHISM AND DIRECT IMAGE

Consider a complex n–dimensional vector bundle

$$p : E \longrightarrow X,$$

denoted by ξ. The total space E is not compact. Denote by ξ^E the one–point compactification of E and call it *the Thom space of the bundle* ξ. This space can be represented a quotient space $\mathbf{D}(\xi)/S(\xi)$ where $\mathbf{D}(\xi)$ is the subbundle of unit balls and $\mathbf{S}(\xi)$ is subbundle of unit spheres. Then we say that the group $K^0(\mathbf{D}(\xi)/\mathbf{S}(\xi)) = K_c(E)$ is *K–group with compact supports* of this noncompact space. This notation can be extended to suspensions. Hence there is a bigraded group

$$K_c^*(E) = K_c^0(E) \oplus K_c^1(E).$$

Consider the element

$$\beta(\xi) \in K^0(\mathbf{D}(\xi), \mathbf{S}(\xi)) \tag{4.18}$$

defined in the chapter 3 (3.11). Then the homomorphism given by multiplication (see (3.12))

$$\beta : K^*(X) \longrightarrow K_c^*(E)$$

is called the Thom homomorphism.

Theorem 42 *The Thom homomorphism is an isomorphism.*

Proof.

Actually Theorem 42 is a consequence of the Bott periodicity Theorem 27. The Bott periodicity theorem is a particular case of Theorem 42 when the bundle ξ is trivial. To reduce Theorem 42 to the Bott periodicity theorem, it is sufficient to consider a pair of base spaces $(Y \subset X)$ such that $X \backslash Y$ has only one cell σ. Then the vector bundle ξ is trivial over σ and also over $\partial \sigma$. Therefore, we can apply Bott periodicity to the pair (X, Y) to obtain:

$$\beta : K^*(X, Y) \longrightarrow K_c^*(E, E_1),$$

where $E_1 = p^{-1}(Y)$. Consider the commutative diagram

$$\cdots \longrightarrow K^*(X, Y) \longrightarrow K^*(X) \longrightarrow K^*(Y) \longrightarrow \cdots$$
$$\Big\downarrow \beta \qquad\qquad \Big\downarrow \beta \qquad\qquad \Big\downarrow \beta$$
$$\cdots \longrightarrow K_c^*(E, E_1) \longrightarrow K_c^*(E) \longrightarrow K_c^*(E_1) \longrightarrow \cdots$$

By induction, two of the vertical homomorphisms are isomorphisms. Therefore, applying the Five Lemma for isomorphisms, we have the proof of the theorem 42. ∎

The Thom isomorphism β can be extended to noncompact bases as a isomorphism of K–groups with compact supports:

$$\beta : K_c^*(X) \longrightarrow K_c^*(E).$$

The homomorphism β can be associated with the inclusion

$$i : X \longrightarrow E$$

of the base X into the total space E as the zero section. Let us denote the homomorphism β by i_*. Consider another vector bundle ψ over the base X. Then there is a commutative diagram of vector bundles:

$$
\begin{array}{ccc}
E_1 & \longleftarrow & E_2 \\
\Big\downarrow \eta & \quad p^{-1}(\eta) \Big\updownarrow j & \\
X & \underset{i}{\overset{\xi}{\rightleftarrows}} & E
\end{array}
$$

where i and j are inclusions as zero sections.

Theorem 43 *The inclusions i and j satisfy the following conditions:*

1. $(ji)_* = j_* i_*,$

2. $i_* \left(x i^*(y) \right) = i_*(x)y$ *for any* $y \in K_c^*(E)$ *and* $x \in K_c^*(X).$

Proof.

Since $y = i_*(z) = \beta z$, we have

$$
\begin{aligned}
i_* \left(x i^*(\beta z) \right) &= \beta i_*(\beta)xz, \\
i_*(x)\beta z &= \beta^2 xz.
\end{aligned}
$$

Hence it is suffices to prove that

$$\beta^2 = \beta i^*(\beta) \tag{4.19}$$

and it is suffices to prove this last equality for the special case when the vector bundle ξ is the one dimensional Hopf bundle η over $\mathbf{CP}^n$. Then the Thom space $\eta^{\mathbf{CP}^n}$ is homeomorphic to $\mathbf{CP}^{n+1}$. In this case, the element β is equal to $\eta - 1$. Let $t \in H^2\left(\mathbf{CP}^n; \mathbf{Q}\right)$ and $\alpha \in H^2\left(\mathbf{CP}^{n+1}; \mathbf{Q}\right)$ be generators. Passing to cohomology via the Chern character in the formula (4.19), we get

$$(\mathrm{ch}\,\beta)^2 = (\mathrm{ch}\,\beta)\left(\mathrm{ch}\,i^*(\beta)\right).$$

Therefore, we need to prove

$$
\left(\alpha + \frac{\alpha^2}{2!} + \ldots + \frac{\alpha^{n+1}}{(n+1)!}\right)\left(\alpha + \frac{\alpha^2}{2!} + \ldots + \frac{\alpha^{n+1}}{(n+1)!}\right) =
$$
$$
= \left(\alpha + \frac{\alpha^2}{2!} + \ldots + \frac{\alpha^{n+1}}{(n+1)!}\right)\left(t + \frac{t^2}{2!} + \ldots + \frac{t^n}{n!}\right). \tag{4.20}
$$

The equality (4.20) follows because

$$
\begin{aligned}
i^*(\alpha) &= t \\
\alpha^k t^l &= \alpha^{k+l}.
\end{aligned}
$$

∎

The Thom isomorphism makes the K–functor into a covariant functor (but not with respect to all continuous mappings), but rather the K–functor can be defined as a covariant functor on the category of smooth manifolds and smooth mappings.

Theorem 44 *For smooth mappings* $f : X \longrightarrow Y$ *of smooth manifolds, there is a homomorphism*

$$f_* : K_c(TX) \longrightarrow K_c(TY)$$

such that

1. *$(fg)_* = f_* g_*$,*
2. *$f_*(x f^*(y)) = f_*(x) y$ for any $y \in K_c^*(TY)$ and $x \in K_c^*(TX)$.*

Proof.

Let $\varphi : X \longrightarrow V$ be an inclusion into a Euclidean space of large dimension. Then the mapping f splits into a composition

$$X \xrightarrow{i} Y \times V \xrightarrow{p} Y, \tag{4.21}$$

where $i(x) = (f(x), \varphi(x))$ and p is the projection onto the first summand. Let $U \subset Y \times V$ be a tubular neighborhood of submanifold $i(X)$. The total space of the tangent bundle TX is a noncompact manifold whose tangent bundle has natural complex structure. Therefore, the inclusion

$$Ti : TX \longrightarrow TU$$

is the inclusion as the zero section of the complexification of the normal bundle of the inclusion $i : X \longrightarrow U$. Let

$$j : TU \longrightarrow TY \times TV$$

be the identity inclusion of an open domain. Let

$$k : TY \longrightarrow TY \times TV$$

be the zero section inclusion.

Definition 10 *Define*

$$f_*(x) k_*^{-1} j_* i_*(x), \quad x \in K_c(TX),$$

where $j_* : K_c^*(TU) \longrightarrow K_c^*(T(Y \times V))$ *is a natural extension of the difference construction from TU to $T(Y \times V)$.*

It is clear that if dim V is sufficiently large then the definition 10 is independent of the choice of the inclusion φ.

The property 1 of Theorem 44 follows from the following commutative diagram of inclusions and projections:

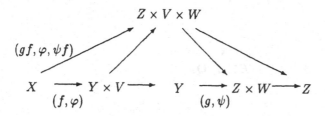

Let us prove the property 2 of Theorem 44. If $x \in K_c^*(TX)$, $y \in K_c^*(TY)$, then

$$f_* \left(x f^*(y) \right) = k_*^{-1} j_* i_* \left(x \left(i^* p^* y \right) \right)$$
$$= k_*^{-1} j_* \left(i_*(x) p^*(y) \right) k_*^{-1} \left((j_* i_*(x)) p^*(y) \right).$$

Since

$$k_*(a) = p_*(a) k_*(1),$$

it follows that

$$j_* i_*(x) = p^* \left(k_*^{-1} j_* i_*(x) \right) k_*(1).$$

Hence

$$k_*^{-1} \left(j_* i_*(x) p^*(y) \right) \left(k_*^{-1} j_* i_*(x) \right) y = f_*(x) y.$$

■

Remark

Theorem 44 can be generalized to the case of almost complex smooth manifolds, that is, manifolds where the tangent bundle admits a complex structure. In particular, Theorem 44 holds for complex analytic manifolds.

4.4   THE RIEMANN–ROCH THEOREM

The Thom isomorphism considered in the section 4.3 is similar to the Thom isomorphism in cohomology. The latter is the operation of multiplication by the element $\varphi(1) \in H_c^{2n}(E; \mathbf{Z})$ where $n = \dim \xi$ is the dimension of the complex vector bundle ξ with total space E.

It is natural to compare these two Thom isomorphisms. Consider the diagram

$$
\begin{array}{ccc}
K^*(X) & \xrightarrow{\;\;\beta\;\;} & K_c^*(E) \\
\downarrow{\scriptstyle \text{ch}} & & \downarrow{\scriptstyle \text{ch}} \\
H^*(X; \mathbf{Q}) & \xrightarrow{\;\;\varphi\;\;} & H_c^*(E; \mathbf{Q})
\end{array}
\tag{4.22}
$$

The diagram (4.22) is not, in general, commutative. In fact,

$$
\begin{aligned}
\text{ch } \beta(x) &= \text{ch } (\beta \otimes x) = \text{ch } \beta\, \text{ch } x, \\
\varphi \text{ch } x &= \varphi(1) \text{ch } x.
\end{aligned}
\tag{4.23}
$$

Therefore, the left hand sides of (4.22) differ from each of other by the factor $T(\xi) = \varphi^{-1}(\text{ch } \beta)$:

$$
\text{ch } \beta(x) = T(\xi)\varphi\,(\text{ch } (x))
$$

The class $T(\xi)$ is a characteristic class and, therefore, can be expressed in terms of symmetric functions. Let $\xi = \xi_1 \oplus \ldots \oplus \xi_n$ be a splitting into one–dimensional summands. Then it is clear that

$$
\beta = \beta_1 \otimes \ldots \otimes \beta_n,
$$

where the β_k the Bott elements of the one dimensional vector bundles. Similarly,

$$
\varphi(1) = \varphi_1(1) \cdot \ldots \cdot \varphi_n(1),
$$

where each $\varphi_k : H^*(X; \mathbf{Z}) \longrightarrow H_c^*(E_k; \mathbf{Z})$ is the Thom isomorphism for the bundle ξ_k. Then

$$
T(\xi) = \varphi^{-1}(\text{ch } \beta) = \varphi_1^{-1}(\text{ch } \beta_1) \cdot \ldots \cdot \varphi_n^{-1}(\text{ch } \beta_n) = T(\xi_1) \cdot \ldots \cdot T(\xi_n).
$$

Hence it is sufficient to calculate $T(\xi)$ for a one dimensional vector bundle, particularly for the Hopf bundle over $\mathbf{CP}^n$. Let $t \in H^2(\mathbf{CP}^n; \mathbf{Z})$ be the generator. Then

$$
T(\xi) = \varphi^{-1}(\text{ch } \beta) = \frac{e^t - 1}{t}.
$$

Thus for an n–dimensional vector bundle ξ we have

$$T(\xi) = \prod_{k=1}^{n} \frac{e^{t_k} - 1}{t_k}.$$

The class $T(\xi)$ is a nonhomogeneous cohomology class which begins with 1. The inverse class $T^{-1}(\xi)$ is called *the Todd class* of the vector bundle ξ.

Theorem 45 *Let $f : X \longrightarrow Y$ be a smooth mapping of smooth manifolds. Then*

$$ch\, f_*(x)T^{-1}(TY) = f_*\left(ch\, xT^{-1}(TX)\right), \quad x \in K_c^*(TX).$$

Proof.

Let

$$TX \xrightarrow{i} U \xrightarrow{j} TY \times \mathbf{R}^{2n} \xrightarrow{p} TY$$

be the decomposition induced by (4.21). Then

$$\begin{aligned}
\mathrm{ch}\, f_*(x) = \mathrm{ch}\, k_*^{-1} j_* i_*(x) &= \\
&= k_*^{-1}\mathrm{ch}\, j_* i_*(x) = k_*^{-1} j_* \mathrm{ch}\, i_*(x) = \\
&= k_*^{-1} j_* \left(i_* \mathrm{ch}\,(x)T(\xi)\right) = f_*\left(\mathrm{ch}\,(x)T(\xi)\right),
\end{aligned}$$

where ξ is the normal bundle to the inclusion i. Then up to a trivial summand we have

$$\xi = -[TX] + f^*[TY].$$

Therefore,

$$T(\xi) = T^{-1}(TX)f^*T(TY)$$

and thus,

$$\mathrm{ch}\, f_*(x) = f_*\left(\frac{f^*T(TY)}{T(TX)}\right). \tag{4.24}$$

Examples

1. Let X be a compact almost complex $2n$–dimensional manifold, Y a one point space, and $f : X \longrightarrow Y$ be the natural mapping. Then the direct image in cohomology

$$f_* : H^*(X; \mathbf{Z}) \longrightarrow H^*(\mathrm{pt}) = \mathbf{Z}$$

maps the group $H^{2n}(X; \mathbf{Z}) = \mathbf{Z}$ isomorphically and the other cohomology groups map to zero. In more detail, we have

$$f_*(x) = \langle x, [X] \rangle \in \mathbf{Z}.$$

Applying the formula (4.24) we get

$$F_*(1) \in K^*(\text{pt}) = \mathbf{Z}.$$

Hence ch $f_*(1)$ is an integer. Therefore, $f_* \left(T^{-1}(TX) \right)$ is also an integer. In other words,

$$\langle T^{-1}(TX), [TX] \rangle \in \mathbf{Z}. \tag{4.25}$$

This number is called the *Todd genus* .

The formula (4.25) gives an integer theorem for almost complex manifolds. In general, the Todd genus is rational number, but in the case of an almost complex manifolds the Todd genus is an integer.

2.   Assume that the n–dimensional almost complex manifold F has Todd genus $\langle T^{-1}(F), [F] \rangle = 1$. Consider a locally trivial bundle $f : E \longrightarrow X$ with fiber F and structure group of the diffeomorphisms which preserve the almost complex structure. Then

$$\text{ch } f_*(x) = f_* \left(\text{ch } x T^{-1}(TF) \right), \ x \in K_c^*(E),$$

where TF means the complex bundle over E of tangent vectors to the fibers. In the case $n = 4$ we have

$$\text{ch } f_*(1) = f_* \left(T^{-1}(TF) \right).$$

The direct image f_* in cohomology decreases the dimension exactly by $2n$. Hence the zero–dimensional component of $f_* \left(T^{-1}(TF) \right)$ equals 1. Therefore, $f_*(1)$ starts with 1 and hence is invertible.

Corollary 7 *The homomorphism*

$$f^* : K^*(X) \longrightarrow K^*(E)$$

is a monomorphism onto a direct summand.

Now we can continue our study of the homomorphism (3.20)

$$b : R(G) \longrightarrow K(BG).$$

from the ring of virtual representations of the compact group G. Let $T \subset G$ be a maximal torus. Then there is the commutative diagram

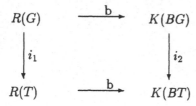

The lower horizontal homomorphism b becomes an isomorphism after completion of the rings. The homomorphism i_1 becomes a monomorphism onto the subring of virtual representations of the torus T which are invariant with respect to action of the Weyl group of inner automorphisms. The homomorphism i_2 also maps the group $K(BG)$ into subring of $K(BT)$ of elements which are invariant with respect to the action of the Weyl group. It is known that the fiber of i_2, G/T, admits an almost complex structure with unit Todd genus (A.Borel, F.Hirzebruch [12]). Then Corollary 7 says that i_2 is a monomorphism. Hence the upper horizontal homomorphism b should be an isomorphism after completion of the rings with respect to the topology induced by inclusion (see M.F.Atiyah, F.Hirzebruch [11]).

5

ELLIPTIC OPERATORS ON SMOOTH MANIFOLDS AND K–THEORY

In this chapter we describe some of the most fruitful applications of vector bundles, namely, in elliptic operator theory. We study some of the geometrical constructions which appear naturally in the analysis of differential and pseudodifferential operators on smooth manifolds.

5.1 SYMBOLS OF PSEUDODIFFERENTIAL OPERATORS

Consider a linear differential operator A which acts on the space of smooth functions of n real variables:

$$A : C^\infty(\mathbf{R}^n) \longrightarrow C^\infty(\mathbf{R}^n).$$

The operator A is a finite linear combination of partial derivatives

$$A = \sum_\alpha a_\alpha(x) \frac{\partial^{|\alpha|}}{\partial x^\alpha}, \qquad (5.1)$$

where the $\alpha = (\alpha_1, \dots, \alpha_n)$ is multi-index, $a_\alpha(x)$ are smooth functions and

$$\frac{\partial^{|\alpha|}}{\partial x^\alpha} = \frac{\partial^{|\alpha|}}{(\partial x^1)^{\alpha_1} (\partial x^2)^{\alpha_2} \dots (\partial x^n)^{\alpha_n}},$$

$$|\alpha| = \alpha_1 + \dots + \alpha_n$$

is the operator given by the partial derivatives.

The maximal value of $|\alpha|$ is called the *order of the differential operator* differential operator , so the formula (5.1) can be written

$$A = \sum_{|\alpha| \le m} a_\alpha(x) \frac{\partial^{|\alpha|}}{\partial x^\alpha},$$

Let us introduce a new set of variables $\xi = (\xi_1, \xi_2, \ldots, \xi_n)$. Put

$$a(x, \xi) = \sum_{|\alpha| \le m} a_\alpha(x) \xi^\alpha i^{|\alpha|},$$

where $\xi^\alpha = \xi_1^{\alpha_1} \xi_2^{\alpha_2} \cdots \xi_n^{\alpha_n}$. The function $a(x, \xi)$ is called the *symbol of a differential operator* A. The operator A can be reconstructed from its symbol by substitution of the operators $\frac{1}{i} \frac{\partial}{\partial x^k}$ for the variables ξ_k, that is,

$$A = a\left(x, \frac{1}{i}\frac{\partial}{\partial x}\right).$$

Since the symbol is polynomial with respect to variables ξ, it can be split into homogeneous summands

$$a(x, \xi) = a_m(x, \xi) + a_{m-1}(x, \xi) + \ldots + a_0(x.\xi).$$

The highest term $a_m(x, x)$ is called the *principal symbol* of the operator A while whole symbol sometimes is called the *full symbol* . The reason for singling out the principal symbol is as follows:

Proposition 13 *Let*

$$y^k = y^k\left(x^1, \ldots, x^n\right) \ (\ or\ y = y(x))$$

be a smooth change of variables. Then in the new coordinate system the operator B *defined by the formula*

$$(Bu)(y) = (Au\left(y(x)\right))_{x=x(y)}$$

is again a differential operator of order m *for which the principal symbol is*

$$b_m(y, \eta) = a_m\left(x(y), \eta \frac{\partial y(x(y))}{\partial x}\right). \tag{5.2}$$

The formula (5.2) shows that variables ξ change as a tensor of valency $(0, 1)$, that is, as components of a cotangent vector.

The concept of a differential operator exists on an arbitrary smooth manifold M. The concept of a whole symbol is not well defined but the principal symbol can be defined as a function on the total space of the cotangent bundle T^*M. It is clear that the differential operator A does not depend on the principal symbol alone but only up to the addition of an operator of smaller order.

The notion of a differential operator can be generalized in various directions. First of all notice that if

$$\mathbf{F}_{x\longrightarrow\xi}(u)(\xi) = \frac{1}{(2\pi)^{n/2}} \int\limits_{\mathbf{R}^n} e^{-i(x,\xi)} u(x)\,dx$$

and

$$\mathbf{F}_{\xi\longrightarrow x}(v)(x) = \frac{1}{(2\pi)^{n/2}} \int\limits_{\mathbf{R}^n} e^{i(x,\xi)} v(\xi)\,d\xi$$

are the direct and inverse Fourier transformations then

$$(Au)\,(x) = \mathbf{F}_{\xi\longrightarrow x}\left(a(x,\xi)\left(\mathbf{F}_{x\longrightarrow\xi}(u)(\xi)\right)\right) \tag{5.3}$$

Hence we can enlarge the family of symbols to include some functions which are not polinomials. Namely, suppose a function a defined on the cotangent bundle T^*M satisfies the condition

$$\left|\frac{\partial^{|\alpha|}}{\partial\xi^\alpha} \frac{\partial^{|\beta|}}{\partial x^\beta} a(x,\xi)\right| \le C_{\alpha,\beta}(1+|\xi|)^{m-|\alpha|} \tag{5.4}$$

for some constants $C_{\alpha,\beta}$. Denote by $\mathbf{S}$ the Schwartz space of functions on $\mathbf{R}^n$ which satisfy the condition

$$\left|x^\beta \frac{\partial^{|\alpha|}}{\partial x^\alpha} u(x)\right| \le C_{\alpha,\beta}$$

for any multiindexes α and β. Then the operator A defined by formula (5.3) is called a *pseudodifferential operator of order m* (more exactly, not greater than m). The pseudodifferential operator A acts in the Schwartz space $\mathbf{S}$.

This definition of a pseudodifferential operator can be extended to the Schwartz space of functions on an arbitrary compact manifold M. Let $\{U_\alpha\}$ be an atlas of charts with a local coordinate system $x_\alpha = (x_\alpha^1, \ldots, x_\alpha^n)$. Without loss of generality we can assume that the local coordinate system x_α maps the chart U_α onto the space $\mathbf{R}^n$. Let $\xi_\alpha = (\xi_{1\alpha}, \ldots, \xi_{n\alpha})$ be the corresponding components of a cotangent vector. Let $\{\varphi_\alpha\}$ be a partition of unity subordinate to the atlas of charts, that is,

$$0 \leq \varphi_\alpha(x) \leq 1, \ \sum_\alpha \varphi_\alpha(x) \equiv 1, \ \mathbf{Supp} \ \varphi_\alpha \subset U_\alpha.$$

Finally, let $\psi_\alpha(x)$ be functions such that

$$\mathbf{Supp} \ \psi_\alpha \subset U_\alpha, \ \varphi_\alpha(x)\psi_\alpha(x) \equiv \varphi_\alpha(x).$$

Then we can define an operator A by the formula

$$A(u)(x) = \sum_\alpha \psi_\alpha(x) A_\alpha \left(\varphi_\alpha(x)u(x) \right), \tag{5.5}$$

where A_α is a pseudodifferential operator on the chart U_α (which is diffeomorphic to $\mathbf{R}^n$) with principal symbol

$$a_\alpha(x_\alpha, \xi_\alpha) = a(x, \xi).$$

When the function $a(x, \xi)$ is polynomial (of order m) the operator A defined by formula (5.5) is a differential operator not depending on the choice of functions ψ_α. In general, the operator A depends on the choice of functions ψ_α, φ_α and the local coordinate system x_α, uniquely up to the addition of a pseudodifferential operator of order strictly less than m.

The next useful generalization consists of a change from functions on the manifold M to smooth sections of vector bundles. Let ξ_1 and ξ_2 be two vector bundles over the manifold M. Consider a linear mapping

$$a : \pi^*(\xi_1) \longrightarrow \pi^* \xi_2, \tag{5.6}$$

where $\pi : T^*M \longrightarrow M$ is the natural projection. Then in any local coordinate system (x_α, ξ_α) the mapping (5.6) defines a matrix valued function which we require to satisfy the condition (5.4). Then the mapping (5.6) defines a pseudodifferential operator

$$A = a(D) : \Gamma^\infty(\xi_1) \longrightarrow \Gamma^\infty(\xi_2)$$

by formulas similar to (5.5), again uniquely up to the addition of a pseudodifferential operator of the order less than m. The crucial property of the definition (5.5) is the following

Proposition 14 *Let*

$$a : \pi^*(\xi_1) \longrightarrow \pi^*(\xi_2), ; \quad b : \pi^*(\xi_a) \longrightarrow \pi^*(\xi_3), ;$$

be two symbols of orders m_1, m_2. Let $c = ba$ be the composition of the symbols. Then the operator

$$b(D)a(D) - c(D) : \Gamma^\infty(\xi_1) \longrightarrow \Gamma^\infty(\xi_3)$$

is a pseudodifferential operator of order $m_1 + m_2 - 1$.

Proposition 14 leads to a way of solving equations of the form

$$Au = f \tag{5.7}$$

for certain pseudodifferential operators A. To find a solution of (5.7), it sufficient to construct a left inverse operator B, that is, $BA = 1$. If B is the pseudodifferential operator $B = b(D)$ then

$$1 = b(D)a(D) = c(D) + (b(D)a(A) - c(D)) .$$

Then by 14, the operator $c(D)$ differs from identity by an operator of order -1. Hence the symbol c has the form

$$c(x, \xi) = 1 + \text{ symbol of order } (-1).$$

Hence for existence of the left inverse operator B, it is necessary that symbol b satisfies the condition

$$a(x, \xi)b(x, \xi) = 1 + \text{ symbol of order } (-1). \tag{5.8}$$

In particular, the condition (5.8) holds if

Condition 1 $a(x, \xi)$ *is invertible for sufficiently large $|\xi| \geq C$.*

In fact, if the condition (5.8) holds then we could put

$$b(x, \xi) = a^{-1}(x, \xi)\chi(x, \xi),$$

where $\chi(x, \xi)$ is a function such that

$$\chi(x, \xi) \equiv 1 \text{ for } |\xi| \geq 2C,$$
$$\chi(x, \xi) \equiv 0 \text{ for } |\xi| \leq C.$$

Then the pseudodifferential operator $a(D)$ is called an *elliptic*   if Condition 1 holds.

The final generalization for elliptic operators is the substitution of a sequence of pseudodifferential operators for a single elliptic operator. Let $\xi_1, \xi_2, \ldots, \xi_k$ be a sequence of vector bundles over the manifold M and let

$$0 \longrightarrow \pi^*(\xi_1) \xrightarrow{a_1} \pi^*(\xi_2) \xrightarrow{a_2} \ldots \xrightarrow{a_{k-1}} \pi^*(\xi_k) \longrightarrow 0 \qquad (5.9)$$

be a sequence of symbols of order $(m_1, \ldots, m_{k-1})$. Suppose the sequence (5.9) forms a complex, that is, $a_s a_{s-1} = 0$. Then the sequence of operators

$$0 \longrightarrow \Gamma^\infty(\xi_1) \xrightarrow{a_1(D)} \Gamma^\infty(\xi_2) \longrightarrow \ldots \longrightarrow \Gamma^\infty(\xi_k) \longrightarrow 0 \qquad (5.10)$$

in general, does not form a complex because we can only know that the composition $a_k(D) a_{k-1}(D)$ is a pseudodifferential operator of the order less then $m_s + m_{s-1}$.

If the sequence of pseudodifferential operators forms a complex and the sequence of symbols (5.9) is exact away from a neighborhood of zero section in T^*M then the sequence (5.10) is called an *elliptic complex* of pseudodifferential operators.

5.2   FREDHOLM OPERATORS

A bounded linear operator

$$F : H \longrightarrow H$$

on a Hilbert space H is called a *Fredholm operator* if

$$\dim \mathbf{Ker}\, F < \infty, \ \dim \mathbf{Coker}\, F < \infty$$

and the image, $\mathbf{Im}\, F$, is closed. The number

$$\text{index}\, F = \dim \mathbf{Ker}\, F - \dim \mathbf{Coker}\, F$$

is called *the index of the Fredholm operator* F. The index can be obtained as

$$\text{index}\, F = \dim \mathbf{Ker}\, F - \dim \mathbf{Ker}\, F^*,$$

where F^* is the adjoint operator.

The bounded operator $K : H \longrightarrow H$ is said to be a *compact*   if any bounded subset $X \subset H$ is mapped to a precompact set, that is, the set $\overline{F(X)}$ is compact.

Theorem 46 *Let F be a Fredholm operator. Then*

1. *there exists $\varepsilon > 0$ such that if $\|F - G\| < \varepsilon$ then G is a Fredholm operator and*
$$index\ F = index\ G,$$

2. *if K is compact then $F + K$ is also Fredholm and*
$$index\ (F + K) = index\ F. \tag{5.11}$$

The operator F is Fredholm if and only if there is an operator G such that both $K = FG - 1$ and $K' = GF - 1$ are compact. If F and G are Fredholm operators then the composition FG is Fredholm and

$$index\ (FG) = index\ F + index\ G.$$

Proof.

Consider first an operator of the form $F = 1 + K$, where K is compact. Let us show that F is Fredholm. The kernel **Ker** F consists of vectors x such that

$$x + Kx = 0, \text{ that is, } Kx = -x.$$

Then the unit ball $B \subset$ **Ker** F is bounded and hence compact, since $K(B) = B$. Hence dim **Ker** F is finite. Let us show that **Im** F is a closed subset. Consider $x_n \in$ **Im** F, $\lim x_n = x$. Then

$$x_n = F(y_n) = y_n + Ky_n.$$

Without loss of generality we can assume that

$$y_n \perp \textbf{Ker}\ F.$$

If the set of numbers $\{\|y_n\|\}$ is unbounded then we can assume that

$$\|y_n\| \longrightarrow \infty.$$

Then

$$\lim_{n \to \infty} \left(\frac{y_n}{\|y_n\|} + K \frac{y_n}{\|y_n\|} \right) = \lim_{n \to \infty} \frac{x}{\|y_n\|} = 0$$

The sequence $y_n / \|y_n\|$ is bounded and by passing to subsequence we can assume that there is a limit

$$\lim_{n \to \infty} K \left(\frac{y_n}{\|y_n\|} \right) = z.$$

Hence

$$\lim_{n \to \infty} \frac{y_n}{\|y_n\|} = -z, \tag{5.12}$$

that is,

$$F(z) = z + Kz = 0.$$

On the other hand, we have that z is orthogonal to **Ker** F, that is, $z = 0$. The latter contradicts (5.12). This means that sequence $\{y_n\}$ is bounded and hence there is a subsequence such that

$$\lim_{n \to \infty} Ky_n = z.$$

Therefore,

$$\lim_{n \to \infty} y_n = \lim_{n \to \infty} (x_n - Ky_n) = x - z.$$

Hence

$$x = \lim_{n \to \infty} x_n = \lim_{n \to \infty} F(y_n) = F(x - z),$$

that is, $x \in \mathbf{Im}\, F$. Thus **Im** F is closed and hence

$$\mathbf{Coker}\, F = \mathbf{Ker}\, F^*.$$

The operator $F^* = 1 + K^*$ has a finite dimensional kernel and, therefore,

$$\dim \mathbf{Coker}\, F = \dim \mathbf{Ker}\, F^* < \infty.$$

This means that F is Fredholm.

Assume that for compact operators K and K'

$$FG = 1 + K, \ GF = 1 + K'.$$

Then we have

$$\mathbf{Ker}\, F \subset \mathbf{Ker}\, (1 + K'),$$

that is, $\dim \mathbf{Ker}\, F < \infty$. Similarly, we have

$$\mathbf{Im}\, F \supset \mathbf{Im}\, (1 + K).$$

Hence the image **Im** F is closed and has finite codimension. This means that both F and G are Fredholm.

Suppose again that F is a Fredholm operator. Consider two splittings of the space H into direct sums:

$$\begin{aligned} H &= \mathbf{Ker}\, F \oplus (\mathbf{Ker}\, F)^\perp \\ H &= (\mathbf{Im}\, F)^\perp \oplus \mathbf{Im}\, F. \end{aligned} \tag{5.13}$$

Then the operator F can be represented as a matrix with respect to the splittings (5.13):

$$F = \begin{Vmatrix} 0 & 0 \\ 0 & F_1 \end{Vmatrix}, \tag{5.14}$$

where F_1 is an isomorphism. We define a new operator G by the following matrix

$$G = \begin{Vmatrix} 0 & 0 \\ 0 & F_1^{-1} \end{Vmatrix},$$

Then the operators

$$K = FG - 1 = \begin{Vmatrix} 1 & 0 \\ 0 & 0 \end{Vmatrix} \text{ and } K' = GF - 1 = \begin{Vmatrix} 1 & 0 \\ 0 & 0 \end{Vmatrix}$$

are finite dimensional and hence compact.

Let $\varepsilon > 0$ be such that $\|G\| < \frac{1}{\varepsilon}$. Suppose that the operator α is such that $\|\alpha\| < \varepsilon$ and $F_1 = F + \alpha$. Notice that both operators $1 + \alpha G$ and $1 + G\alpha$ are invertible. Then

$$(1 + G\alpha)^{-1} G F_1 = (1 + G\alpha)^{-1}(1 + D\alpha + K) = 1 + (1 + G\alpha)^{-1} K,$$
$$F_1 G (1 + \alpha G)^{-1} = (1 + \alpha G + K')(1 + \alpha G)^{-1} = 1 + K'(1 + \alpha G)^{-1}.$$

Since operators $(1 + G\alpha)^{-1} K$ and $K'(1 + \alpha G)^{-1}$ are compact, it follows that the operator F' is Fredholm.

Let us estimate the index of the Fredholm operator. If the operator F has matrix form

$$F = \begin{Vmatrix} F_1 & 0 \\ 0 & F_2 \end{Vmatrix} \tag{5.15}$$

then, evidently,

$$\text{index } F = \text{index } F_1 + \text{index } F_2.$$

If both Hilbert spaces H_1 and H_2 are finite dimensional and $F : H_1 \longrightarrow H_2$ is bounded then F is Fredholm and

$$\text{index } F = \dim H_1 + \dim H_2.$$

If F is invertible then

$$\text{index } F = 0.$$

If in the operator (5.15), $F_1 = 0$ is finite dimensional and F_2 is invertible then F is Fredholm and index $F = 0$.

Hence, by adding a compact summand we can change the Fredholm operator so that the index does not change and a given finite dimensional subspace is included in the kernel of the resulting operator.

Suppose that F has the splitting (5.14) and let G be another operator such that $\|F - G\| < \varepsilon$. Let

$$G = \left\| \begin{matrix} \alpha_{11} & \alpha_{12} \\ \alpha_{21} & F_2 + \alpha_{22} \end{matrix} \right\|, \tag{5.16}$$

where $\|\alpha_{22}\| < \varepsilon$. For sufficiently small α the operator $F_2 + \alpha_{22}$ is invertible. Then for $X = -(F_2 + \alpha_{22})^{-1}\alpha_{21}$ we have

$$G = \left\| \begin{matrix} 1 & 0 \\ X & 0 \end{matrix} \right\| \left\| \begin{matrix} \beta_{11} & \beta_{12} \\ 0 & F_2 + \beta_{22} \end{matrix} \right\|.$$

Similarly, for $Y = -\beta_{12}(F_2 + \beta_{22})^{-1}$,

$$G = \left\| \begin{matrix} 1 & 0 \\ X & 1 \end{matrix} \right\| \left\| \begin{matrix} \gamma_{11} & 0 \\ 0 & F_2 + \gamma_{22} \end{matrix} \right\| \left\| \begin{matrix} 1 & Y \\ 0 & 1 \end{matrix} \right\|.$$

Thus

$$\text{index } G = \text{index } F.$$

Now let us prove (5.11). We can represent the operators F and K as a matrices

$$F = \left\| \begin{matrix} F_1 & 0 \\ 0 & F_2 \end{matrix} \right\|; \quad K = \left\| \begin{matrix} K_{11} & K_{12} \\ K_{21} & K_{22} \end{matrix} \right\|$$

such that F_1 is finite dimensional and K_{22} is sufficiently small. Then we are in situation of the condition (5.16). Hence

$$\text{index } (F + K) = \text{index } F.$$

∎

The notion of a Fredholm operator has an interpretation in terms of the finite dimensional homology groups of a complex of Hilbert spaces. In general, consider a sequence of Hilbert spaces and bounded operators

$$0 \longrightarrow C_0 \overset{d_0}{\longrightarrow} C_1 \overset{d_1}{\longrightarrow} \ldots \overset{d_{n-1}}{\longrightarrow} C_n \longrightarrow 0. \tag{5.17}$$

We say that the sequence (5.17) is *Fredholm complex* if $d_k d_{k-1} 0$, $\text{Im } d_k$ is a closed subspace and

$$\dim (\textbf{Ker } d_k / \textbf{Coker } d_{k-1}) = \dim H (C_k, d_k) < \infty.$$

Then the *index of Fredholm complex* (5.17) is defined by the following formula:

$$\text{index}\,(C,d) = \sum_k (-1)^k \dim H(C_k, d_k).$$

Theorem 47 *Let*

$$0 \longrightarrow C_0 \overset{d_0}{\longrightarrow} C_1 \overset{d_1}{\longrightarrow} \ldots \overset{d_{n-1}}{\longrightarrow} C_n \longrightarrow 0 \qquad (5.18)$$

be a sequence satisfying the condition that each $d_k d_{k-1}$ is compact. Then the following conditions are equivalent:

1. *There exist operators $f_k : C_k \longrightarrow C_{k-1}$ such that $f_{k+1} d_k + d_{k-1} f_k = 1 + r_k$ where each r_k is compact.*

2. *There exist compact operators s_k such that the sequence of operators $d'_k = d_k + s_k$ forms a Fredholm complex. The index of this Fredholm complex is independent of the operators s_k.*

We leave the proof to the reader. Theorem 47 allows us to generalize the notion of a Fredholm complex using one of the equivalent conditions from Theorem 47.

Exercise

Consider the Fredholm complex (5.18). Form the new short complex

$$C_{ev} \;=\; \bigoplus_k C_{2k},$$

$$C_{odd} \;=\; \bigoplus_k C_{2k+1},$$

$$D \;=\; \begin{Vmatrix} d_0 & d_1^* & 0 & \ldots & 0 \\ 0 & d_2 & d_3^* & \ldots & 0 \\ 0 & 0 & d_4 & \ldots & 0 \\ & \ldots & & \ddots & \end{Vmatrix} \qquad (5.19)$$

Prove that the complex (5.19) is Fredholm with the same index.

5.3   THE SOBOLEV NORMS

Consider the Schwartz space **S**. Define the *Sobolev norm* by the formula

$$\|u\|_s^2 = \int_{\mathbf{R}^n} \bar{u}(x)(1+\Delta)^s u(x)dx,$$

where

$$\Delta = \sum_{k=1}^{n}\left(i\frac{\partial}{\partial x^k}\right)^2$$

is the Laplace operator. Using the Fourier transformation this becomes

$$\|u\|_s^2 = \int_{\mathbf{R}^n}(1+|\xi|^2)^s|\hat{u}(\xi)|^2 d\xi \tag{5.20}$$

where the index s need not be an integer.

The Sobolev space $H_s(\mathbf{R}^n)$ is the completion of **S** *with respect to the Sobolev norm (5.20).*

Proposition 15 *The Sobolev norms (5.20) for different coordinate systems are all equivalent on any compact $K \subset \mathbf{R}^n$, that is, there are two constants $C_1 > 0$ and $C_2 > 0$ such that for* **Supp** $u \subset K$

$$\|u\|_{1,s} \leq C_1\|u\|_{2,s} \leq C_2\|u\|_{1,s}.$$

Proposition 15 can easily be checked when s is an integer. For arbitrary s we leave it to the reader. This proposition allows us to generalize the Sobolev spaces to an arbitrary compact manifold M and vector bundle ξ. Let $\{U_\alpha\}$ be an atlas of charts where the vector bundle is trivial over each U_α. Let $\{\varphi_\alpha\}$ be a partition of unity subordinate to the atlas. Let $u \in \Gamma^\infty(M, \xi)$ be a section. Put

$$\|u\|_s^2 = \sum_\alpha \|\varphi_\alpha u\|_s^2. \tag{5.21}$$

Proposition 15 says that the definition (5.21) defines a Sobolev norm, well defined up to equivalent norms. Hence the completion of the space of sections $\Gamma^\infty(M, \xi)$ does not depend on the choice of partition of unity or the choice of local coordinate system in each chart U_α. We shall denote this completion by $H_s(M, \xi)$.

Theorem 48 *Let M be a compact manifold, ξ be a vector bundle over M and $s_1 < s_2$. Then the natural inclusion*

$$H_{s_2}(M,\xi) \longrightarrow H_{s_1}(M,\xi) \tag{5.22}$$

is a compact operator.

Theorem 48 is called *the Sobolev inclusion theorem.*

Proof.

For the proof it is sufficient to work in a local chart since any section $u \in \Gamma^\infty(M,\xi)$ can be split into a sum

$$u = \sum_\alpha \varphi_\alpha u,$$

where each summand $\varphi_\alpha u$ has support in the chart U_α. So let u be a function defined on $\mathbf{R}^n$ with support in the unit cube $\mathbf{I}^n$. These functions can be considered as functions on the torus $\mathbf{T}^n$. Then a function u can be expanded in a convergent Fourier series

$$u(x) = \sum_l a_l e^{i(l,x)}.$$

Partial differentiation transforms to multiplication of each coefficient a_l by the number l_k, where l is a multiindex $l = (l_1, \ldots, l_n)$. Therefore,

$$\|u\|_s^2 = \sum_l |a_l|^2 \left(1 + |l_1|^2 + \ldots + |l_n|^2\right)^s . \tag{5.23}$$

By formula (5.23), the space $H_s(\mathbf{T}^n)$ is isomorphic to the Hilbert space l_2 of the square summable sequences by the correspondence

$$u(x) \mapsto \left\{ b_l = \frac{a_l}{(1 + |l_1|^2 + \ldots + |l_n|^2)^{s/2}} \right\}$$

Then the inclusion (5.22) becomes an operator $l_2 \longrightarrow l_2$ defined by the correspondence

$$\{b_l\} \mapsto \left\{ \frac{b_l}{(1 + |l_1|^2 + \ldots + |l_n|^2)^{(s_2 - s_1)/2}} \right\}$$

which is clearly compact.

Theorem 49 *Let*
$$a(D) : \Gamma^\infty(M, \xi_1) \longrightarrow \Gamma^\infty(M, \xi_2) \qquad (5.24)$$
be a pseudodifferential operator of order m. Then there is a constant C such that
$$\|a(D)u\|_{s-m} \leq C\|u\|_s, \qquad (5.25)$$
that is, the operator $a(D)$ can be extended to a bounded operator on Sobolev spaces:
$$a(D) : H_s(M, \xi_1) \longrightarrow H_{s-m}(M, \xi_2). \qquad (5.26)$$

Proof.

The theorem is clear for differential operators. Indeed, it is sufficient to obtain estimates in a chart, as the symbol σ has compact support. If $a(D) = \frac{\partial}{\partial x^j}$ then

$$\|\frac{\partial}{\partial x^j} u(x)\|_s^2 = \int_{\mathbf{R}^n} (1 + |\xi|^2)^s \left| \widehat{\frac{\partial}{\partial x^j} u} \right|^2 d\xi =$$

$$= \int_{\mathbf{R}^n} (1 + |\xi|^2)^s |\xi_j \hat{u}(\xi)|^2 d\xi \leq \int_{\mathbf{R}^n} (1 + |\xi|^2)^{s+1} |\hat{u}(\xi)|^2 d\xi =$$

$$= \|u(x)\|_{s+1}^2. \qquad (5.27)$$

Hence the inequality (5.25) follows from (5.27) by induction.

For pseudodifferential operators, the required inequality can also be obtained locally using a more complicated technique.   ∎

Using theorems 48 and 49 it can be shown that an elliptic operator is Fredholm for appropriate choices of Sobolev spaces.

Theorem 50 *Let $a(D)$ be an elliptic pseudodifferential operator of order m as in (5.24). Then its extension (5.26) is Fredholm. The index of the operator (5.26) is independent of the choice of the number s.*

Proof.

As in section 5.1 we can construct a new symbol b of order $-m$ such that both $a(D)b(D) - 1$ and $b(D)a(D) - 1$ are pseudodifferential operators of order -1. Hence by Theorem 46, $a(D)$ gives a Fredholm operator (5.26).

To prove that the index of $a(D)$ does not depend of the number s, consider the special operator $(1+\Delta)^k$ with symbol $(1 + |\xi|^2)^k$. Since the norm $\|(1+\Delta)^k u\|_s$ is equivalent to the norm $\|u\|_{s+2k}$, the operator

$$(1 + \Delta)^k : H_{s+2k}(\xi) \longrightarrow H_s(\xi)$$

is an isomorphism. Then the operator

$$A = (1 + \Delta)^{-k} \sigma(D)(1 + \Delta)^k : H_{s+2k}(\xi_1) \longrightarrow H_s(\xi_1) \longrightarrow$$
$$\longrightarrow H_{s-m}(\xi_2) \longrightarrow H_{s+2k-m}(\xi_2)$$

differs from $\sigma(D)$ by a compact operator and therefore has the same index. ∎

Corollary 8 *The kernel of an elliptic operator $\sigma(D)$ consists of infinitely smooth sections.*

Proof.

Indeed, by increasing the number s we have a commutative diagram

$$
\begin{array}{ccc}
H_s(\xi_1) & \xrightarrow{\ \sigma(D)\ } & H_{s-m}(\xi_2) \\
\uparrow{\scriptstyle U} & & \uparrow{\scriptstyle U} \\
H_{s+1}(\xi_1) & \xrightarrow{\ \sigma(D)\ } & H_{s+1-m}(\xi_2)
\end{array}
\qquad (5.28)
$$

An increase in the number s can only decrease the dimension of the kernel, **Ker** $\sigma(D)$. Similarly, the dimension of the cokernel may only decrease since the cokernel is isomorphic to the kernel of the adjoint operator $\sigma(D)^*$. Since the index does not change, the dimension of kernel cannot change. Hence the kernel does not change in the diagram (5.28). Thus the kernel belongs to the $\bigcap_s H_s(\xi_1) = \Gamma^\infty(\xi_1)$.

5.4   THE ATIYAH–SINGER FORMULA FOR THE INDEX OF AN ELLIPTIC OPERATOR

In the sense explained in previous sections, an elliptic operator $\sigma(D)$ is defined by a symbol

$$\sigma : \pi^*(\xi_1) \longrightarrow \pi^*(\xi_2) \tag{5.29}$$

which is an isomorphism away from a neighborhood of the zero section of the cotangent bundle T^*M. Since M is a compact manifold, the symbol (5.29) defines a triple $(\pi^*(\xi_1), \sigma, \pi^*(\xi_2))$ which in turn defines an element

$$[\sigma] \in K_c(T^*M).$$

Theorem 51 *The index* index $\sigma(D)$ *of the Fredholm operator* $\sigma(D)$ *depends only on the element* $[\sigma] \in K_c(T^*M)$.

The mapping

$$index : K_c(T^*M) \longrightarrow \mathbf{Z}$$

is an additive homomorphism.

Proof.

A homotopy of an elliptic symbol gives a homotopy of Fredholm operators and, under homotopy, the index does not change. Assume that the symbol σ is an isomorphism not only away from zero section but everywhere. The operator $\sigma(D)$ can be decomposed into a composition of an invertible operator $(1 + \Delta)^{m/2}$ and a operator $\sigma_1(D)$ of the order 0. The symbol σ_1 is again invertible everywhere and is therefore homotopic to a symbol σ_2 which is independent of the cotangent vector and also invertible everywhere. Then the operator $\sigma_2(D)$ is multiplication by the invertible function σ_2. Therefore, $\sigma_2(D)$ is invertible. Thus

$$index\ \sigma(D) = 0.$$

Finally, if $\sigma = \sigma_1 \oplus \sigma_2$ then $\sigma(D) = \sigma_1(D) \oplus \sigma_2(D)$ and hence

$$index\ \sigma(D) = index\ \sigma_1(D) + index\ \sigma_2(D).$$

The total space of the cotangent bundle T^*M has a natural almost complex structure. Therefore, the trivial mapping $p : M \longrightarrow \text{pt}$ induces the direct image homomorphism

$$p_* : K_c(T^*M) \longrightarrow K_c(\text{pt}) = \mathbf{Z},$$

as described in Section 4.3

Theorem 52 *Let $\sigma(D)$ be an elliptic pseudodifferential operator. Then*

$$index\ \sigma(D) = p_*[\sigma]. \tag{5.30}$$

Using the Riemann-Roch theorem 45, the formula (5.30) can be written as

$$\text{index } \sigma(D) = \big(\text{ch } [\sigma]T^{-1}(T^*M), [T^*M]\big)$$

where $[T^*M]$ is the fundamental (open) cycle of the manifold T^*M. For an oriented manifold M we have the Thom isomorphism

$$\varphi : H^*(M) \longrightarrow H_c^*(T^*M). \tag{5.31}$$

Therefore, the formula (5.31) has the form

$$\text{index } \sigma(D) = \big(\varphi^{-1}\text{ch } [\sigma]T^{-1}(T^*M), [M]\big) \tag{5.32}$$

The formula (5.32) was proved by M.F.Atiyah and I.M.Singer [9] and is known as the *Atiyah–Singer formula.*

Proof.

The proof of the Atiyah–Singer formula is technically complicated. Known proofs are based on studying the algebraic and topological properties of both the left and right hand sides of (5.30). In particular, Theorem 51 shows that both the left and right hand sides of the (5.30) are homomorphisms on the group $K_c(T^*M)$.

Let $j : M_1 \longrightarrow M_2$ be a smooth inclusion of compact manifolds and let σ_1 be an elliptic symbol on the manifold M_1. Assume that σ_2 is a symbol on the manifold M_2 such that

$$[\sigma_2] = j_*[\sigma_1]. \tag{5.33}$$

Theorem 53 *If two symbols σ_1 and σ_2 satisfy the condition (5.33) then the elliptic operators $\sigma_1(D)$ and $\sigma_2(D)$ have the same index,*

$$\text{index } \sigma_1(D) = \text{index } \sigma_2(D).$$

Theorem 52 follows from Theorem 53.

In fact, Theorem 52 holds for $M = \text{pt}$. Then the inclusion

$$j : \text{pt} \longrightarrow \mathbf{S}^n.$$

gives the direct image

$$j_* : K_c(\text{pt}) \longrightarrow K_c\left(T^*\mathbf{S}^n, T^*_{s_0}\right)$$

which is an isomorphism. Therefore, for a symbol σ defining a class

$$[\sigma] \in K_c\left(T^*\mathbf{S}^n, T^*_{s_0}\right)$$

we have

$$[\sigma] = j_*[\sigma'], \ [\sigma'] = p_*[\sigma],$$
$$\text{index } \sigma(D) = \text{index } \sigma'(D) = [\sigma'] = p_*([\sigma]).$$

Finally, let $j : M \longrightarrow \mathbf{S}^n$ an inclusion and $q : \mathbf{S}^n \longrightarrow \text{pt}$ be the natural projection. Then

$$p = qj, \ j_*([\sigma]) = [\sigma'],$$

and by Theorem 53

$$\text{index } \sigma(D) = \text{index } \sigma'(D) = q_*[\sigma'] = q_*j_*[\sigma]p_*[\sigma].$$

Thus the Atiyah–Singer formula follows from Theorem 53.

Let us explain Theorem 53 for the example when the normal bundle of the inclusion $M_1 \longrightarrow M_2$ is trivial. The symbol $[\sigma_2] = j_*[\sigma_1]$ is invertible outside of a neighborhood of submanifold $M_1 \subset T^*M_2$. Suppose that the symbols are of order 0. Then the symbol σ_2 can be chosen so that it is independent of the cotangent vector outside of a neighborhood U of submanifold M_1. Then we can chose the operator $\sigma_2(D)$ such that if $\mathbf{Supp} \cap U = \emptyset$ then $\sigma_2(D)(u)$ is multiplication by a function. Hence we can substitute for the manifold M_2, the Cartesian product $M_1 \times T^k$, where T^k is torus, equipped with an elliptic operator of the same index. Now we can use induction with respect to the integer k and that means that it is sufficient consider the case when $k = 1$.

The problem is now reduced to the existence of an elliptic operator $\tau(D)$ on the circle such that

$$\text{index } \tau(D) = 1$$

and

$$[\tau] \in K_c \left(T^* \mathbf{S}^1, \, T^*_{s_0} \right)$$

is a generator.

6

SOME APPLICATIONS OF VECTOR BUNDLE THEORY

Here we describe some problems where vector bundles appear in a natural way. We do not pretend to give a complete list but rather they correspond to the author's interest. Nevertheless, we hope that these examples will demonstrate the usefulness of vector bundle theory as a geometric technique.

6.1   SIGNATURES OF MANIFOLDS

The *signature* of a compact oriented $4n$-dimensional manifold X is the signature of a quadratic form on the space $H^{2n}(X; \mathbf{Q})$ of $2n$-dimensional rational cohomology given by the formula

$$(x, y) = \langle xy, [X] \rangle, \qquad (6.1)$$

where $x, y \in H^{2n}(X; \mathbf{Q})$, $[X]$ is fundamental $4n$-dimensional cycle and $\langle , \rangle$ is the natural value of a cocycle on a cycle.

By Poincaré duality we see that the quadratic form (6.1) is nondegenerate. This form can be reduced to a form with a diagonal matrix with respect to a basis of the vector space $H^{2n}(X; \mathbf{Q})$. Notice that the signature of the quadratic form is the same as the signature of a similar quadratic form on the space $H^{2n}(X; \mathbf{R})$. This allows us to describe the cohomology groups using de Rham differential forms.

The notion of signature can be generalized to a bilinear form on the whole cohomology group $H^*(X) = \bigoplus_{k=0}^{4n} H^k(X)$ using the same definition (6.1). This

215

is also a nondegenerate form but it is not symmetric. Nevertheless, consider the splitting

$$H^*(X) = H^{ev}(X) \oplus H^{odd}(X).$$

Then the bilinear form (6.1) splits into a direct sum of a symmetric form on $H^{ev}(X)$ and a skewsymmetric form on $H^{odd}(X)$. Let us agree that the signature of a skewsymmetric form equals zero. Then the signature of the manifold X is equal to the signature of the bilinear form (6.1) on the $H^*(X)$. In fact, the form (6.1) decomposes into a direct sum of the subspaces

$$H^*(X) = H^{odd}(X) \oplus \left(H^0(X) \oplus H^{4n}(X) \right) \oplus$$
$$\oplus \left(H^2(X) \oplus H^{4n-2}(X) \right) \oplus \ldots \oplus \left(H^{2n-2}(X) \oplus H^{2n+2}(X) \right) \oplus H^{2n}(X).$$

The matrix of the bilinear form then has the form

$$\left\| \begin{matrix} 0 & A \\ A^t & 0 \end{matrix} \right\| \tag{6.2}$$

on each summand except the first and last. The signature of the matrix (6.2) is trivial. Hence the signature of the form on the H^* coincides with the signature on the $H^{2n}(X)$.

Theorem 54 *If a compact oriented $4n$–dimensional manifold X is the boundary of an oriented compact $4n + 1$–dimensional manifold Y then*

$$sign\ X = 0.$$

Proof.

The homology group (with rational or real coefficients) can be thought of as

$$H_k(X) = \hom \left(H^k(X);\ \mathbf{R} \right).$$

Then the bilinear form (6.1) can be considered as an isomorphism

$$D : H^k(X) \longrightarrow H_{4n-k}(X)$$

coinciding with $\cap$–product with the fundamental cycle $[X] \in H_{4n}(X)$ inducing Poincaré duality. The manifold Y also has a Poincaré duality induced by $\cap$–product with the fundamental cycle $[Y] \in H_{4k+1}(Y, \partial Y)$:

$$D' = \cap[Y] : \quad H^k(Y) \longrightarrow H_{4n+1-k}(Y, X),$$
$$D'' = \cap[Y] : \quad H^k(Y, X) \longrightarrow H_{4n+1-k}(Y).$$

The exact homology sequence for the pair (Y, X) fits into a commutative diagram

$$\cdots \longrightarrow H^k(Y,X) \xrightarrow{\;j^*\;} H^k(Y) \xrightarrow{\;i^*\;} H^k(X) \xrightarrow{\;\partial\;} H^{k+1}(Y,X) \longrightarrow \cdots$$

$$\left\downarrow D'' \qquad\qquad \left\downarrow D' \qquad\qquad \left\downarrow D \qquad\qquad \left\downarrow D''$$

$$\cdots \longrightarrow H_{4n+1-k}(Y) \xrightarrow{\;j^*\;} H_{4n+1-k}(Y,X) \xrightarrow{\;\delta\;} H_{4n-k}(X) \xrightarrow{\;i^*\;} H_{4n-k}(Y) \longrightarrow \cdots \tag{6.3}$$

Let $k = 2n$. Then the dimension of the image $\mathbf{Im}\, i^* \subset H^{2n}(X)$ can be calculated as an alternating sum of dimensions of all spaces which lie on the left side of upper line of the diagram (6.3):

$$\dim \mathbf{Im}\, i^* = \dim H^{2n}(Y) - \dim H^{2n}(Y,X) + \dim H^{2n-1}(X) - $$
$$- \dim H^{2n-1}(Y) + \dim H^{2n-1}(Y,X) - \dim H^{2n-2}(X) + \ldots$$

Using the right lower part of (6.3) we have

$$\mathrm{codim}\, \mathbf{Im}\, i_* = \mathrm{codim}\, \mathbf{Im}\, \delta = \mathrm{codim}\, \mathbf{Ker}\, i_* =$$
$$= \dim H_{2n}(Y) - \dim H_{2n}(Y,X) + \dim H_{2n-1}(X) -$$
$$- \dim H_{2n-1}(Y) + \dim H_{2n-1}(Y,X) - \dim H_{2n-2}(X) + \ldots$$

Since $\dim H^k(X) = \dim H_k(X)$,

$$\dim \mathbf{Im}\, i^* = \mathrm{codim}\, \mathbf{Im}\, i^* = \frac{1}{2} \dim H^{2n}(X).$$

Then the group $H^{2n}(X)$ decomposes into two summands:

$$H^{2n}(X) = A \oplus B, \; A = \mathbf{Im}\, i^*, B = (\mathbf{Ker}\, i_*)^*$$

By commutativity of the diagram, $D(A) \subset B^*$ and hence

$$D = \left\| \begin{matrix} 0 & D_2 \\ D_3 & D_4 \end{matrix} \right\|, \; D_3 = D_2^t.$$

Thus the signature of D is trivial. ∎

Theorem 54 shows that the signature of the manifold X depends only on the bordism class and hence can be expressed in terms of characteristic classes of the manifold X. This means that there should exist a polynomial $L(X)$ of the Pontryagin classes such that

$$\text{sign}\, X = \langle L(X), [X] \rangle.$$

This class $L(X)$ is called the *Hirzebruch–Pontryagin class*.

6.1.1 Calculation of the Hirzebruch–Pontryagin class

The Hirzebruch–Pontryagin class and the signature can be given an interpretation as the index of an elliptic operator on the manifold X. Such operator exists and is called the *Hirzebruch operator*.

To construct the Hirzebruch operator, consider the de Rham complex of differential forms on the manifold X:

$$0 \longrightarrow \Omega_0 \xrightarrow{\ d\ } \Omega_1 \xrightarrow{\ d\ } \ldots \xrightarrow{\ d\ } \Omega_{4n} \longrightarrow 0 \qquad (6.4)$$

The space Ω_k is the space of sections of the vector bundle Λ_k over X, the exterior power of the cotangent bundle $T^*X = \Lambda_1$. The complex (6.4) is an elliptic differential complex for which the symbols are given by exterior multiplication by the cotangent vector $\xi \in T^*X$.

Notice that the signature of the manifold X is defined in terms of multiplication of differential forms

$$(\omega_1, \omega_2) = \int_X \omega_1 \wedge \omega_2.$$

Fix a Riemannian metric g on the manifold X. The metric g induces a scalar product on the vector bundles Λ_k. Let $e_1, \ldots, e_{4k}$ be an orthogonal basis in the fiber of T^*X. Then the corresponding orthogonal basis in the fiber of Λ_k consists of the vectors $e_{i_1} \wedge \ldots \wedge e_{i_k}$. Consider the mapping

$$* : \Lambda_k \longrightarrow \Lambda_{4n-k},$$

which to each basis vector $e_{i_1} \wedge \ldots \wedge e_{i_k}$ associates the vector $(-1)_\sigma e_{j_1} \wedge \ldots \wedge e_{j_{4n-k}}$ where the e_{j_l} complement the system of vectors $e_{i_1}, \ldots, e_{i_k}$ to form the basis and σ is the parity of the permutation $(i_1, \ldots, i_k, j_1, \ldots, j_{4n-k})$. Then $*^2 = (-1)^k$ on Λ_k. Thus we have a homomorphism

$$* : \Gamma(\Lambda_k) \longrightarrow \Gamma(\Lambda_{4n-k}).$$

Let μ be the differential form which defines the measure on the manifold X induced by the Riemannian metric. In terms of the local coordinate system,

$$\mu = \sqrt{\det g_{ij}} \, dx^1 \wedge \ldots \wedge dx^{4n}.$$

It is clear that $\mu = *(1)$ and for any function f we have $*(f\omega) = f * (\omega)$.

Denote by $\langle \omega_1, \omega_2 \rangle$ the function whose value is equal to the scalar product of the forms ω_1 and ω_2. Then

$$\langle \omega_1, \omega_2 \rangle \mu = \omega_1 \wedge *\omega_2 = (-1)^k * \omega_1 \wedge \omega_2. \tag{6.5}$$

In the space $\bigoplus_k \Gamma \Lambda_k$ we have the positive scalar product

$$\langle \theta_1, \omega_2 \rangle = \int \langle \omega_1, \omega_2 \rangle \mu. \tag{6.6}$$

Then taking into account (6.5),

$$(\omega_1, \omega_2) = \langle \omega_1, *\omega_2 \rangle \tag{6.7}$$

for $\omega_1 \in \Gamma(\Lambda_k)$, $\omega_2 \in \Gamma\Lambda_{4n-k})$.

Let d^* denote the adjoint operator with respect to the scalar product (6.6). This means that

$$\langle \omega_1, d\omega_2 \rangle \equiv \langle d^*\omega_1, \omega_2 \rangle.$$

Using (6.7),

$$\langle d^*\omega_1, \omega_2 \rangle = \langle \omega_1, d\omega_2 \rangle = \int_X \omega_1 \wedge *d\omega_2 =$$

$$(-1)^k \int_X *\omega_1 \wedge \omega_2 = \int_X d(*\omega_1 \wedge \omega_2) - \int_X d * \omega_1 \wedge \omega_2 =$$

$$-\int_X d * \omega_1 \wedge \omega_2 = -\int_X *d * \omega_1 \wedge *\omega_2 = -\langle *d * \omega_1, \omega_2 \rangle.$$

Hence

$$d^* = - * d * . \tag{6.8}$$

Now we can identify the de Rham cohomology with the kernel of the Laplace operator

$$\Delta = (d + d^*)^2 = d^*d + dd^*.$$

Note that the Laplace operator commutes with both d and d^*.

Theorem 55 *The operator $D = d + d^*$ is an elliptic differential operator and in the space $\Gamma^\infty(\Lambda)$*

$$\mathbf{Ker}\ d = \mathbf{Ker}\ D \oplus \mathbf{Im}\ d.$$

Proof.

The first statement of Theorem 55 follows directly from the previous chapter since the symbol of the operator D is an invertible matrix away from zero section of the cotangent bundle.

Let $\omega \in \Gamma^\infty(\Lambda_k)$. Then $\omega \in H_s(\Lambda_*)$. Decompose the element ω into a sum

$$\omega = \omega_1 + \omega_2,$$

where

$$\omega_1 \in \mathbf{Ker}\ D \subset H_s(\Lambda_*),\ \omega_2 \in (\mathbf{Ker}\ D)^\perp.$$

On the other hand,

$$(D(H_{s+1}(\Lambda_*)))^\perp = \mathbf{Ker}\ D$$

Hence

$$(\mathbf{Ker}\ D)^\perp = D(H_{s+1}(\Lambda_*)),$$

that is,

$$\omega_2 = D(\Omega),\ \Omega \in H_{s+1}(\Lambda_*).$$

Let $s' > s$ and

$$\omega = \omega_1' + \omega_2',\ \omega_1' \in \mathbf{Ker}\ D \subset H_{s'}(\Lambda_*),\ \omega_2' \in (\mathbf{Ker}\ D)^\perp.$$

Then

$$\omega_1 - \omega_1' = \omega_2 - \omega_2' \in (\mathbf{Ker}\ D)^\perp.$$

Hence $\omega_1 = \omega_1'$, $\omega_2 = \omega_2'$ and $\omega_2 \in \Gamma^\infty(\Lambda_*)$. If $\omega \in \mathbf{Ker}\ d$ then

$$\omega = \omega_1 + D\omega_2, \omega_1 \in \mathbf{Ker}\ D,$$

that is,

$$0 = d\xi\omega = d\omega_1 + dD\omega_2.$$

Hence

$$0 = dD\omega_2 = Dd^*\omega_2, \; dd^*\omega_2 = 0, \; d^*\omega_2 = 0.$$

Thus

$$\omega = \omega_1 + d^*\omega_2.$$

∎

Theorem 55 says that cohomology group $H^k(X)$ is isomorphic to **Ker** D in the space $\Gamma^\infty(\Lambda_k)$. Notice that by formula (6.8), the operator $*$ preserves the kernel **Ker** D.

Consider a new operator

$$\alpha = i^{k(k-1)+2n} * : \Lambda_k \longrightarrow \Lambda_{4n-k}.$$

Then

$$\alpha^2 = 1, \; d^* = -\alpha d\alpha, \; \alpha D = -D\alpha.$$

Thus there is a splitting

$$\Lambda_* = \Lambda^+ \oplus \Lambda^-$$

into the two eigen subspaces corresponding to the eigenvalues $+1$ and -1 of the operator α. Then

$$D\left(\Gamma^\infty\left(\Lambda^+\right)\right) \subset \Gamma^\infty\left(\Lambda^-\right),$$
$$D\left(\Gamma^\infty\left(\Lambda^-\right)\right) \subset \Gamma^\infty\left(\Lambda^+\right).$$

Hence

$$\textbf{Ker } D = \textbf{Ker } D^+ \oplus \textbf{Ker } D^-,$$

where D^+ and D^- are restrictions of D to $\Gamma^\infty(\Lambda_+)$ and $\Gamma^\infty(\Lambda_-)$. Thus

$$\text{sign } X = \dim \textbf{Ker } D^+ - \dim \textbf{Ker } D^-.$$

Since the operator D is selfadjoint, the operator D^+ is adjoint to the operator D^-. Hence

$$\text{sign } X = \text{index } D^+.$$

The operator D^+ is called *the Hirzebruch operator*. To calculate the index of the Hirzebruch operator, let Σ be the symbol of the Hirzebruch operator. The symbol Σ is a homomorphism

$$\Sigma : \pi^*\left(c\Lambda^+\right) \longrightarrow \pi^*\left(c\lambda^-\right), \tag{6.9}$$

which is an isomorphism away from the zero section of the cotangent bundle. Hence

$$\text{index } D^+ = \langle \text{ch } [\Sigma] T(X), [T^* X] \rangle,$$

where $T(X)$ is the Todd class of the complexification of the tangent bundle.

The construction of the element $[\Sigma]$ defined by the triple (6.9) can be extended to arbitrary orientable real bundles over the base X. Indeed, let $p : E \longrightarrow X$ be the projection on the total space of the bundle ξ. Let

$$0 \longrightarrow p^* \Lambda_0(c\xi) \xrightarrow{\sigma} p^* \Lambda_1(c\xi) \xrightarrow{\sigma} \ldots \xrightarrow{\sigma} p^* \Lambda_n(c\xi) \longrightarrow 0$$

be the complex of vector bundles where σ is the exterior multiplication by the vector $\xi \in E$. Similar to the tangent bundle, the bundle $\Lambda_*(\xi) = \oplus \Lambda_k(\xi)$ has an operation

$$* : \Lambda_k(\sigma) \longrightarrow \Lambda_{n-k}(\xi)$$

which induces a splitting

$$\Lambda_*(c\xi) = \Lambda^+(\xi) \oplus \Lambda^-(\xi)$$

and a homomorphism

$$\Sigma = \sigma - *\sigma*$$

mapping

$$\Sigma : \Lambda^+(\xi) \longrightarrow \Lambda^-(\xi) \tag{6.10}$$

isomorphically away from the zero section of the total space E. Hence the triple (6.10) defines an element $[\Sigma] \in K_c(E)$. Notice that if $X = X_1 \times X_2$ and $\xi = \xi_1 \oplus \xi_2$, where ξ_k are vector bundles over X_k, then

$$[\Sigma(\xi)] = [\Sigma \xi_1] \otimes [\Sigma \xi_2].$$

Then the problem can be reduced to the universal two dimensional vector bundle ξ over $\mathbf{BSO}(2) = \mathbf{BU}(1) = \mathbf{CP}^\infty$. In this case, the bundle ξ is isomorphic to the complex Hopf bundle and $c\xi = \xi \oplus \xi^*$. Hence

$$\Lambda^+(\xi) = \overline{\mathbf{1}} \oplus \xi^*$$
$$\Lambda^-(\xi) = \xi \oplus \overline{\mathbf{1}}.$$

Then the symbol $\Sigma : \Lambda^+(\xi) \longrightarrow \Lambda^-(\xi)$ has the following form

$$\Sigma = \left\| \begin{matrix} z & 0 \\ 0 & z \end{matrix} \right\|,$$

where z is a vector in the fiber of the bundle ξ. Hence

$$[\Sigma] = ([\eta] - 1) \oplus (1 - [\eta^*]) = [\eta] - [\eta_*],$$

where η is the Hopf bundle over the Thom complex $\xi^{\mathbf{CP}^\infty} = \mathbf{CP}^\infty$. Hence

$$\operatorname{ch} [\Sigma] = e^\alpha - e^{-\alpha}, \; \alpha \in H_2 (\mathbf{CP}^\infty; \mathbf{Z}).$$

Thus if the bundle ξ is decomposed into a direct sum of two dimensional bundles

$$\xi = \xi_1 \oplus \cdots \oplus \xi_n,$$

then

$$\operatorname{ch} (\Sigma(\xi)) = \prod_{k=1}^{n} (e^{\alpha_k} - e^{-\alpha_k}) = \varphi \left(\prod_{k=1}^{n} \frac{e^{\alpha_k} - e^{-\alpha_k}}{\alpha_k} \right),$$

where $\varphi : H^*(X) \longrightarrow H_c^* (\xi^X)$ is the Thom isomorphism. Then

$$\varphi^{-1} (\operatorname{ch} (\Sigma)T(X)) = \prod_{k=1}^{n} \frac{\alpha_k}{e^{\alpha_k} - 1} \frac{(-\alpha_k)}{e^{-\alpha_k} - 1} \frac{e^{\alpha_k} - a^{-\alpha_k}}{\alpha_k} =$$

$$= 2^n \prod_{k=1}^{n} \frac{\alpha_k/2}{\tanh(\alpha_k/2)}.$$

Thus for a $4n$–dimensional manifold X,

$$L(X) = 2^{2n} \prod_{k=1}^{n} \frac{t_k/2}{\tanh(t_k/2)},$$

where the t_k are formal generators for the tangent bundle of the manifold X.

6.2 C^*–ALGEBRAS AND K–THEORY

The space $\Gamma(\xi)$ of sections of a vector bundle ξ over a base B is a module over the ring $\mathbf{C}(B)$ of continuous functions on the base B. The ring $\mathbf{C}(B)$ also can be considered as a section space $\Gamma(\overline{1})$ of the one dimensional trivial vector bundle $\overline{1}$. Hence $\Gamma(\overline{1})$ is a one dimensional free module. Hence for an arbitrary vector bundle ξ, the module $\Gamma(\xi)$ is a finitely generated projective module.

Theorem 56 *The correspondence $\xi \mapsto \Gamma(\xi)$ is an equivalence between the category of complex vector bundles over a compact base X and the category of finitely generated projective modules over the algebra $\mathbf{C}(B)$.*

Proof.

If two vector bundles ξ and η are isomorphic then the modules $\Gamma(\xi)$ and $\Gamma(\eta)$ of sections are isomorphic. Therefore, it is sufficient to prove that any finitely generated projective module P over the C^*-algebra $C(B)$ is isomorphic to a module $\Gamma(\xi)$ for a vector bundle ξ, and any homomorphism

$$\varphi : \Gamma(\xi) \longrightarrow \Gamma(\eta) \tag{6.11}$$

is induced by a homomorphism of bundles

$$f : \xi \longrightarrow \eta.$$

We start by proving the second statement. Let (6.11) be a homomorphism of modules. Assume first that both vector bundles ξ and η are trivial, that is,

$$\xi = B \times \mathbf{C}^n, \ \eta = B \times \mathbf{C}^m.$$

Let

$$a_1, \ldots, a_n \in \mathbf{C}^n$$
$$b_1, \ldots, b_m \in \mathbf{C}^m$$

be bases. The vectors a_k, b_k induces constant sections

$$\bar{a}_k \in \Gamma(\xi), \ \bar{b}_k \in \Gamma(\eta).$$

Hence any sections $\sigma \in \Gamma(\xi)$, $\tau \in \Gamma(\eta)$ decompose into linear combinations

$$\sigma = \sum_k \sigma^k(x)\bar{a}_k, \qquad \sigma^k(x) \in C(B),$$

$$\tau = \sum_l \tau^l \bar{b}_l, \qquad \tau^l(x) \in C(B).$$

Therefore, the homomorphism φ has the following form

$$\varphi(\sigma) = \sum_k \sigma_k(x)\varphi(\bar{a}_k) = \sum_k \sigma^k(x) \left(\sum_l \varphi_k^l(x)\bar{b}_l \right) =$$

$$= \sum_{k,l} \sigma^k(x)\varphi_k^l(x)\bar{b}_l.$$

The matrix valued function $\varphi_k^l(x)$ defines a fiberwise homomorphism of bundles

$$f = \|\varphi_k^l(x)\| : \xi \longrightarrow \eta. \tag{6.12}$$

Clearly, the homomorphism (6.12) induces the initial homomorphism (6.11).

Now let ξ and η be arbitrary vector bundles. Complement them to form trivial bundles, that is,

$$\xi \oplus \xi_1 = \bar{n}, \; \eta \oplus \eta_1 = \bar{m}.$$

Let

$$p : \bar{n} \longrightarrow \bar{n}, \; q : \bar{m} \longrightarrow \bar{m}$$

be projections onto the first summands ξ and η, respectively. Define a new homomorphism

$$\psi : \Gamma(\bar{n}) \longrightarrow \Gamma(\bar{m})$$

by the matrix

$$\psi = \left\| \begin{matrix} \varphi & 0 \\ 0 & 0 \end{matrix} \right\|.$$

It is clear that

$$\psi\pi = \kappa\psi,$$

where $\pi : \Gamma(\bar{n})$ and $\kappa : \Gamma(\bar{m})$ are homomorphisms induced be p and q. We see that ψ is induced by a homomorphism of trivial vector bundles

$$g : \bar{n} \longrightarrow \bar{m}.$$

Hence the homomorphism g has the matrix form

$$g = \left\| \begin{matrix} f & 0 \\ 0 & 0 \end{matrix} \right\|,$$

where $f = qgp$ is the homomorphism from the bundle ξ to the vector bundle η which induces the homomorphism (6.11).

Now we prove the first statement. Let P be a finitely generated projective module over the algebra $C(B)$. By definition the module P is the direct summand of the free module $C(B)^n = \Gamma(\bar{n})$

$$P \oplus Q = \Gamma(\bar{n}).$$

The module P is the image of a projection

$$\varphi : \Gamma(\bar{n}) \longrightarrow \Gamma(\bar{n}), \; \varphi^2 = \varphi, \; P = \mathbf{Im} \; \varphi.$$

The projection φ, a homomorphism of modules, is induced by a homomorphism g of the trivial vector bundle $\bar{n}$. Hence $g^2 = g$, that is, g is a fiberwise projection. Therefore, the image of g is a locally trivial vector bundle,

$$\bar{n} = \xi \oplus \eta, \; \xi = \mathbf{Im} \; g.$$

Thus $P = \Gamma(\xi)$.   ■

Theorem 56 says that the group $K(B)$ is naturally isomorphic to the group $K_0(C(B))$ in the algebraic K–theory for the ring $C(B)$:

$$K(B) \longrightarrow K_0(C(B)). \tag{6.13}$$

We can generalize the situation considered in Theorem 56 to the case of a Cartesian product $B = X \times Y$. Let ξ be a vector bundle over a base B. Fix a point $y_0 \in Y$. Then the section space $\Gamma(\xi, (X \times y_0))$ is a projective module over the ring $C(X)$. Thus we have a family of projective modules parametrized by points of Y. This leads us to the definition of a vector bundle over a base Y with fiber a projective module $P = \Gamma(\xi, (X \times y_0))$ and structure group the group of automorphisms of the module P.

More generally, let A be an arbitrary C^*–algebra and P a projective A-module. Then a locally trivial bundle over the base Y with the structure group $G = \mathrm{Aut}_A(P)$ will be called a *vector A–bundle*. The corresponding Grothendieck group will be denoted by $K_A(X)$. If $X = \mathrm{pt}$ is a one point space then the group $K_A(\mathrm{pt})$ coincides with $K_0(A)$ in algebraic K–theory. If $B = X \times Y$ then there is a natural isomorphism

$$K(X \times Y) \longrightarrow K_{C(Y)}(X). \tag{6.14}$$

If Y is a one point space then the isomorphism (6.14) coincides with the isomorphism (6.13) from Theorem 56.

Theorem 56 can also be generalized to the case of vector A–bundles. Namely, if ξ is a vector A–bundle over a base Y, then the section space $\Gamma(\xi)$ is a projective module over the algebra $C(Y, A)$ of continuous functions on the Y with values in A. Hence there is a natural isomorphism

$$K_A(Y) \longrightarrow K_0(C(Y, A)).$$

When $B = X \times Y$,

$$C(X \times Y) = C(Y, C(X))$$

and there is a commutative diagram

$$
\begin{array}{ccc}
K(X \times Y) & \longrightarrow & K_0(C(X \times Y)) \\
\downarrow & & \parallel \\
K_{C(X)}(Y) & \longrightarrow & K_0(C(Y, C(X)))
\end{array}
$$

The groups $K_A(X)$ satisfy many properties similar to those of classical complex K–theory. The tensor product induces a natural homomorphism

$$K_A(X) \otimes K(Y) \longrightarrow K_A(X \times Y).$$

In particular, multiplication by the Bott element $\beta \in K^0(S^2, s_0)$ defines an isomorphism

$$K_A(X, Y) \xrightarrow{\otimes \beta} K_A(X \times D^2, (X \times S^1) \cup (Y \times D^2)),$$

and an isomorphism

$$K_A^n(X, Y) \xrightarrow{\otimes \beta} K_A^{n-2}(X, Y).$$

Consider the homomorphism

$$\varphi : K_A(\mathrm{pt}) \otimes K(X) \longrightarrow K_A(X),$$

induced by tensor multiplication. After tensoring with the rational number field Q, there is an isomorphism

$$\varphi^{-1} : K_A^*(X) \otimes \mathbf{Q} = K^*(X) \otimes K_A^*(\mathrm{pt}) \otimes \mathbf{Q}.$$

Define

$$\mathrm{ch}_A = (\mathbf{Id} \otimes \mathrm{ch}) \varphi^{-1} : K_A^*(X) \longrightarrow H^*(X; K_A^*(\mathrm{pt}) \otimes \mathbf{Q}) \qquad (6.15)$$

The homomorphism (6.15) is an analogue of the Chern character and fits into the following commutative diagram

$$
\begin{array}{ccc}
K_A^*(X) \otimes K^*(X) & \xrightarrow{\otimes} & K_A^*(X) \\
\downarrow{\scriptstyle \mathrm{ch}_A \otimes \mathrm{ch}} & & \downarrow{\scriptstyle \mathrm{ch}_A} \\
H^*(X; K_A^*(\mathrm{pt}) \otimes \mathbf{Q}) \otimes H^*(X; \mathbf{Q}) & \xrightarrow{\otimes} & H^*(X; K_A^*(\mathrm{pt}) \otimes \mathbf{Q})
\end{array}
$$

Example

The Calkin algebra.

The Calkin algebra is a quotient algebra $C(\mathcal{H}) = B(H)/\mathrm{Comp}(H)$ where $B(H)$ is the algebra of bounded operators on the Hilbert space H and $\mathrm{Comp}(H)$ is the ideal of compact operators. To study the group $K_{C(\mathcal{H})}(X)$ it is sufficient to consider locally trivial bundles with free $C(\mathcal{H})$–module as fibers. In this case the structure group is $\mathbf{GL}(n, \mathcal{C}(\mathcal{H}))$ — the group of invertible matrices with entries in $\mathcal{C}(\mathcal{H})$. The algebra $\mathrm{End}(\mathcal{C}(\mathcal{H})^n)$ of n–dimensional matrices is isomorphic to the algebra $\mathcal{C}(\mathcal{H})$ since the infinite dimensional Hilbert space H is isomorphic to the direct sum of n copies of H. Hence

$$\mathbf{GL}(n, \mathcal{C}(\mathcal{H})) \sim \mathbf{GL}(1, \mathcal{C}(\mathcal{H})) = \mathbf{GL}(\mathcal{C}(\mathcal{H})).$$

Let

$$\pi : B(H) \longrightarrow \mathcal{C}(\mathcal{H})$$

be the natural projection. Then the inverse image $\pi^{-1}(\mathbf{GL}(\mathcal{C}(\mathcal{H}))$ coincides with the set $\mathcal{F}$ of Fredholm operators. The projection

$$\pi : \mathcal{F} \longrightarrow \mathbf{GL}(\mathcal{C}(\mathcal{H}))$$

is a fibration with the fiber $\mathrm{Comp}(H)$ which satisfies the covering homotopy axiom.

The space $\mathcal{F}$ of Fredholm operators is the union of an infinite number of connected components

$$\mathcal{F} = \bigcup_k \mathcal{F}_k,$$

where $\mathcal{F}_k$ is the space of Fredholm operators of index $= k$. Hence the group $\mathbf{GL}(\mathcal{C}(\mathcal{H}))$ also splits into a union of connected components

$$\mathbf{GL}(\mathcal{C}(\mathcal{H})) = \bigcup_k \mathbf{GL}(\mathcal{C}(\mathcal{H}))_k.$$

Consider the group $\mathbf{GL}(H)$ of all invertible operators on the Hilbert space H and the map

$$\pi : \mathbf{GL}(H) \longrightarrow \mathbf{GL}(\mathcal{C}(\mathcal{H}))_0.$$

This map is clearly an isomorphism. The kernel consists of invertible operators of the form $1 + K$ where K is compact and is denoted by $1 + \mathrm{Comp}$. The group $1 + \mathrm{Comp}$ is the closure of the invertible finite dimensional operators. Each finite dimensional invertible matrix F can be extended to an operator

$$\tilde{F} = \begin{pmatrix} F & 0 \\ 0 & 1 \end{pmatrix} \in 1 + \mathrm{Comp}.$$

Hence there is an inclusion

$$\mathbf{GL}(n, \mathbf{C}) \subset 1 + \mathrm{Comp},$$

and an inclusion of the direct limit

$$\mathbf{GL}(\infty, \mathbf{C}) \subset 1 + \text{Comp}. \qquad (6.16)$$

The inclusion (6.16) is a weak homotopy equivalence. Hence classical K–theory can be represented by locally trivial bundles with the structure group 1+Comp.

Theorem 57 *There is a natural isomorphism*

$$f : K^{-1}_{C(\mathcal{H})}(X, x_0) \longrightarrow K^0(X, x_0).$$

Proof.

Consider the suspension of X as a union of two cones over X. Then any vector bundle over the algebra $C(\mathcal{H})$ can be described by two charts — two copies of the cone. Hence the vector bundle is defined by a single transition function

$$F : X \longrightarrow \mathcal{F}_0.$$

Let $\{U_\alpha\}$ be an atlas of sufficiently small charts on X. Then there are functions

$$\varphi_\alpha : U_\alpha \longrightarrow \text{Comp}(H) \qquad (6.17)$$

such that for any α and $x \in U_\alpha$ the operator $F(x) + \varphi_\alpha(x)$ is invertible. Put

$$\psi_{\alpha\beta}(x) = (F(x) + \varphi_\alpha(x))\,(F(x) + \varphi_\beta(x))^{-1} \in 1 + \text{Comp}. \qquad (6.18)$$

The functions (6.18) define transition functions with values in the structure group $1 + \text{Comp}$. Hence the transition functions (6.18) define an element of $K^0(X)$. This element is independent of the choice of functions (6.17). Therefore, we have a well defined homomorphism

$$f : K^{-1}_{C(\mathcal{H})}(X, x_0) \longrightarrow K^0(X, x_0). \qquad (6.19)$$

The homomorphism (6.19) is a monomorphism. Indeed, if the transition functions (6.18) define a trivial bundle then there are functions

$$u_\alpha : U_\alpha \longrightarrow 1 + \text{Comp}$$

such that

$$\psi_{\alpha\beta}(x) = u_\alpha^{-1}(x) u_\beta(x).$$

Using (6.18),

$$u_\alpha^{-1}(x)\,(F(x) + \varphi_\alpha(x)) = u_\beta^{-1}(x)\,(F(x) + \varphi_\beta(x)) = \Phi(x). \qquad (6.20)$$

Hence the function defined by (6.20) is invertible and differs from $F(x)$ by a compact summand. Since the group $\mathbf{GL}(H)$ is contractible, there is a homotopy $\Phi(x,t)$, $x \in X$, $t \in [0,1]$ such that

$$\Phi(x,1) = \Phi(x), \ \Phi(x,0) \equiv 1. \tag{6.21}$$

Thus the function (6.21) is defined on the cone on X and can serve on the boundary as the transition function on the suspension SX.

Now let us prove that (6.19) is surjective. Consider a system of the transition functions $\psi_{\alpha\beta} : U_{\alpha\beta} \longrightarrow 1 + \mathrm{Comp}$. Using the contractibility of the group $\mathbf{GL}(H)$, we can find a system of functions $h_\alpha : U_\alpha \longrightarrow 1 + \mathrm{Comp}$ such that

$$\psi_{\alpha\beta}(x) = h_\alpha(x)h_\beta^{-1}(x).$$

This means that

$$h_\alpha(x) = h_\beta(x) + \mathrm{Comp}.$$

Hence there is a Fredholm operator valued function $F(x)$ such that

$$F(x) - h_\alpha(x) \in \mathrm{Comp}.$$

There is another interpretation of the homomorphism (6.19).

Theorem 58 *For any continuous map* $F : X \longrightarrow \mathcal{F}$ *there is a homotopic map* $F' : X \longrightarrow \mathcal{F}$ *such that*

$$\dim \mathbf{Ker}\ F(x) \equiv \mathit{Const}, \ \dim \mathbf{Coker}\ F(x) \equiv \mathit{Const}.$$

The unions $\bigcup \mathbf{Ker}\ F'(x)$ *and* $\bigcup \mathbf{Coker}\ F'(x)$ *form locally trivial vector bundles* $[\mathbf{Ker}\ F']$ *and* $[\mathbf{Coker}\ F']$. *The element*

$$\mathit{index}\ F = [\mathbf{Ker}\ F'] - [\mathbf{Coker}\ F'] \in K(X) \tag{6.22}$$

is independent of the choice of F'. *If* $\mathit{index}\ F = 0$ *then the map*

$$\mathit{index}\ : [X, \mathcal{F}] \longrightarrow K(X)$$

defined by (6.22) coincides with (6.19).

6.3 FAMILIES OF ELLIPTIC OPERATORS

Consider a continuous family of Fredholm operators

$$F_x : H \longrightarrow H, \ x \in X,$$

parametrized by a compact space X. Then index F_x is a locally constant function and hence gives no new invariant. But if $\dim \ker F_x \equiv \text{const}$, the kernels $\ker F_x$ form a vector bundle **Ker** F. Similarly, **Coker** F is a vector bundle. In this case, the index of the family of Fredholm operators may be interpreted as an element

$$\text{index } F \overset{\text{def}}{=} [\textbf{Ker } F] - [\textbf{Coker } F] \in K(X).$$

Clearly, the classical index gives

$$\dim \text{index } F = \dim[\textbf{Ker } F] - \dim[\textbf{Coker } F].$$

The generalized index of a family of elliptic operators $\sigma_x(D)$,

$$\text{index } \sigma(D) \in K(X)$$

can be expressed in terms of the symbol of the operator $\sigma(D)$. Let

$$p : M \longrightarrow X$$

be a locally trivial bundle for which fiber is a manifold Y. Let $T_Y^* M$ be the vector bundle annihilating the tangent vectors to the fibers and

$$\pi : T_Y^* M \longrightarrow M$$

be the correspondent projection.

Let ξ_1, ξ_2 be two vector bundles over M. Consider the infinite dimensional vector bundles $p_!(\xi_1)$ and $p_!(\xi_2)$ over X with fibers $\Gamma\left(\xi_1, p^{-1}(x)\right)$ and $\Gamma\left(\xi_2, p^{-1}(x)\right)$, respectively, over the point $x \in X$. Notice that $p^{-1}(x)$ is diffeomorphic to manifold Y. Let $p_!(\xi_i)_s$ denote the bundle associated to $p_!(\xi_i)$ by substituting for the fiber $\Gamma\left(\xi_i, p^{-1}(x)\right)$ its completion by s-Sobolev norm. Then a homomorphism of bundles

$$\sigma(D) : p_!(\xi_1) \longrightarrow p_!(\xi_2) \tag{6.23}$$

is said to be a family of pseudodifferential operators if at each point $x \in X$ the homomorphism of fibers is a pseudodifferential operator of order m on the manifold $p^{-1}(x) \cong Y$. Assume that the symbols of these operators together give a continuous homomorphism

$$\sigma : \pi^*(\xi_1) \longrightarrow \pi^*(\xi_2). \tag{6.24}$$

and that the homomorphism (6.24) is an isomorphism away from the zero section of the bundle $T_Y^*(M)$. In this case we say that the family (6.23) is an elliptic family of pseudodifferential operators. Then, if (6.23) is an elliptic family, it defines both a continuous family of Fredholm operators

$$\sigma(D) : p_!(\xi_1)_s \longrightarrow p_!(\xi_2)_{s-m}$$

and, using (6.24), an element

$$[\sigma] \in K_c\left(T_Y^*(M)\right).$$

Theorem 59 *For an elliptic family (6.23),*

$$index\ \sigma(D) = p_![\sigma] \in K(X) \otimes \mathbf{Q}.$$

Proof.

Let us prove this theorem for the case when $M = X \times Y$. Then there is a second projection

$$q : M \longrightarrow Y.$$

The crucial observation is that

$$\Gamma(\xi, M) = \Gamma\left(q_!(\xi), Y\right) = \Gamma\left(p_!(\xi), X\right).$$

The family of pseudodifferential operators

$$\sigma(D) : p_!(\xi_1) = \Gamma(\xi_1, Y) \times X \longrightarrow \Gamma(\xi_2, Y) \times X \qquad (6.25)$$

generates the homomorphism

$$\Gamma(\sigma(D)) : \Gamma\left(\Gamma(\xi_1, Y), X\right) = \Gamma(\xi_1, X \times Y) \longrightarrow \Gamma(\xi_2, X \times Y) = \Gamma\left(\Gamma(\xi_1, Y), X\right)$$

or

$$q_!(\sigma(D)) : \Gamma\left(\Gamma(\xi_1, X), Y\right) = \Gamma\left(q_!(\xi_1), Y\right) \longrightarrow \Gamma\left(q_!(\xi_2), Y\right) = \Gamma\left(\Gamma(\xi_1, X), Y\right). \qquad (6.26)$$

The last representation (6.26) of the family (6.25) is a pseudodifferential operator over the C^*-algebra $C(X)$.

In the case where the family of elliptic operators has both constant dimensional kernel and cokernel, both the kernel and cokernel of the operator (6.26) over algebra $C(X)$ are projective modules since they can be identified with spaces of sections of finite dimensional vector bundles:

$$\textbf{Ker } q_!(\sigma(D)) = \Gamma\left(\textbf{Ker } \sigma(D)\right)$$

$$\textbf{Coker } q_!(\sigma(D)) = \Gamma\left(\textbf{Coker } \sigma(D)\right)$$

Hence the index of the family of elliptic operators $\sigma(D)$ as an element of the group $K(X)$ is identical to the index of the elliptic operator $q_!(\sigma(D))$ as an element of the group $K_{C(X)}(pt) = K(X)$.

On the other hand, the symbol σ of the family (6.25) generates a homomorphism of vector bundles

$$q_!(\sigma) : \pi^*(q_!(\xi_1)) \longrightarrow \pi^*(q_!(\xi_2))$$

which is the symbol of a pseudodifferential operator (6.26) over C^*-algebra $C(X)$. Hence, in the elliptic case, the element $[q_!(\sigma)] \in K_{C(X),c}(T^*Y)$ is identified with $[\sigma]$ by the isomorphism

$$K_c(T^*Y \times X) \cong K_{C(X),c}(T^*Y).$$

Thus to prove the theorem it is sufficient to generalize the Atiyah–Singer formula to the case of an elliptic operator over the C^*-algebra $C(X)$. This project can be realized (see for example [30]). Firstly, we need to investigate the category of vector bundles over C^*-algebras and study the category of Hilbert modules and so called Fredholm operators over C^*-algebras. We shall follow here ideas of Paschke [35] (see also Rieffel [41]).

Let A be an arbitrary C^*-algebra (with unit). Then a *Hilbert A–module* is a Banach space M which is also a (unital) A-module with a sequilinear form with values in A . Assume that

1. $(x, y) = (y, x)^* \in A, \quad x, y \in M$;

2. $(x, x) \geq 0, \quad x \in M$;

3. $(\lambda x, y) = \lambda(x, y), \quad x, y \in M, \quad \lambda \in A$;

4. $\|x\|^2 = \|(x, x)\|, \quad x \in M$.

A model (infinite dimensional) Hilbert A–module $l_2(A)$ can constructed as follows. By definition, a point $x \in l_2(A)$ is a sequence

$$x = (x_1, x_2, \ldots, x_n, \ldots), \quad x_n \in A$$

such that the series $\sum_{k=1}^{\infty} x_k x_k^*$ converges in norm in the algebra A, that is, there is a limit

$$\|x\|^2 = \lim_{N \longrightarrow \infty} \sum_{k=1}^{N} x_k x_k^* \in A.$$

We define a 'scalar' product by

$$(x, y) \overset{def}{=} \sum_{k=1}^{\infty} x_k y_k^* \in A. \tag{6.27}$$

The following inequality

$$\left\| \sum_k x_k y_k^* \right\|^2 \leq \left\| \sum_k x_k x_k^* \right\| \cdot \left\| \sum_k y_k y_k^* \right\|.$$

implies that the series on the right hand side of (6.27) converges. The space $l_2(A)$ has a decreasing sequence of subspaces

$$[l_2(A)]_n = \{x : x_1 = x_2 = \ldots = x_n = 0\}.$$

A bounded A–linear operator

$$K : l_2(A) \longrightarrow l_2(A)$$

is said to be A–*compact* if

$$\lim_{n \longrightarrow \infty} \|K|_{[l_2(A)]_n}\| = 0.$$

A bounded A–linear operator

$$F : l_2(A) \longrightarrow l_2(A)$$

is said to be A–*Fredholm* if there exists a bounded A–linear operator

$$G : l_2(A) \longrightarrow l_2(A)$$

such that both $FG - 1$ and $GF - 1$ are A–compact.

This terminology is justified by the following properties:

1. Any finitely generated projective A–module admits the structure of a Hilbert module.

2. Let ξ be a A–bundle over a compact manifold X with fiber a finitely generated projective A–module. Then the Sobolev completion $H^s(\xi)$ of the section space is isomorphic to a direct summand of $l_2(A)$.

3. A pseudodifferential A–operator $\sigma(D)$ of order m that maps from the section space $\Gamma(\xi_1)$ to the section space $\Gamma(\xi_2)$ can be extended to a bounded operator of the Sobolev spaces

$$\sigma(D) : H_s(\xi_1) \longrightarrow H^{s-m}(\xi_2). \tag{6.28}$$

4. The natural inclusion

$$H^{s+1}(\xi) \hookrightarrow H^s(\xi)$$

is an A–compact operator.

5. If the operator (6.28) is elliptic then it is A–Fredholm.

6. If F is an A–Fredholm operator then it defines a homotopy invariant

$$\text{index } F \in K_0(A).$$

In fact, for such an operator F,

$$F : H_1 = l_2(A) \longrightarrow H_2 = l_2(A),$$

there exist decompositions

$$H_1 = V_{10} \oplus V_{11}, \quad H_2 = V_{20} \oplus V_{21},$$

such that the corresponding matrix of the operator F has the form

$$F = \left\| \begin{matrix} F_0 & 0 \\ 0 & F_1 \end{matrix} \right\|$$

where F_1 is an isomorphism, the modules V_{10} and V_{20} are finitely generated projective modules and

$$\text{index } F = [V_{10}] - [V_{20}] \in K_0(A).$$

7. If $\sigma(A)$ is an elliptic pseudodifferential A–operator on a compact manifold X then

$$\text{index } \sigma(D) = p_*([\sigma]) \in K_0 \otimes \mathbf{Q},$$

where

$$p : X \longrightarrow \text{pt}$$

and

$$p_* : K_{A,c}(T^*X) \longrightarrow K_A(\text{pt}) = K_0(A)$$

is the direct image in K_A–theory.

The proofs of these properties are very similar to those for classical pseudodifferential operators.

6.4   FREDHOLM REPRESENTATIONS AND ASYMPTOTIC REPRESENTATIONS OF DISCRETE GROUPS

Let

$$\rho : G \longrightarrow A$$

be a unitary representation of a group G in a C^*-algebra A. Then by section 3.4 there is an associated vector bundle $b(\rho) \in K_A(BG)$ giving a homomorphism

$$b : R(G; \, A) \longrightarrow K_A(BG), \tag{6.29}$$

where $R(G; \, A)$ is the ring of virtual representations of the group G in the algebra A.

When A is the algebra of all bounded operators on an infinite dimensional Hilbert space H we do not obtain an interesting homomorphism (6.29): the homomorphism is trivial since all infinite dimensional vector bundles are trivial. For finite dimensional representations it has been shown that, in some sense, the homomorphism (6.29) is an isomorphism when the group G is compact. But in the case of a discrete group such as the fundamental group of a manifold this homomorphism is usually trivial. Therefore, the problem arises to find new examples of geometric representations which give richer information about the group $K(BG)$.

6.4.1   Fredholm representations

There is a so called relative version of representation theory associated with a homomorphism

$$\varphi : A_1 \longrightarrow A_2.$$

We use the term *relative representation* for a triple $\rho = (\rho_1, F, \rho_2)$ where

$$\rho_1, \, \rho_2 : G \longrightarrow A_1$$

are two representations and F is an operator interlacing the representations $\varphi\rho_1$ and $\varphi\rho_2$. In other words, F is an invertible element in A_2 such that

$$F^{-1}\varphi\rho_1(g)F = \varphi\rho_2(g), \quad g \in G. \tag{6.30}$$

The representations ρ_1 and ρ_2 define vector A_1–bundles over BG, $\xi_1 = b(\rho_1)$ and $\xi_2 = b(\rho_2)$. The associated A_2–bundles $\xi_1 \otimes_\varphi A_2$ and $\xi_2 \otimes_\varphi A_2$ are isomorphic:

$$b(F) : \xi_1 \otimes_\varphi A_1 \longrightarrow \xi_2 \otimes_\varphi A_2.$$

The triple $(\xi_1, b(F), \xi_2)$ defines an element of the relative K–theory

$$b(\rho) = (\xi_1, b(F), \xi_2) \in K_\varphi(BG). \tag{6.31}$$

On the other hand, if $\rho_1 = \rho_2$ then the condition (6.30) means that the triple (ρ_1, F, ρ_2) defines a classical representation $\bar\rho$ of the group $G \times \mathbf{Z}$ into the algebra A_2:

$$
\begin{aligned}
\bar\rho(g) &= \varphi\rho(g) \in A_2, \ g \in G; \\
\bar\rho(a) &= F \in A_2, \ \text{where } a \in \mathbf{Z} \text{ is the generator.}
\end{aligned}
$$

Hence the representation $\bar\rho$ defines an element

$$b(\bar\rho) \in K_{A_2}(B(G \times \mathbf{Z})) = K_{A_2}(BG \times \mathbf{S}^1). \tag{6.32}$$

The two elements (6.31) and (6.32) are connected in the classical exact sequence for the homomorphism $\varphi : A_1 \longrightarrow A_2$:

$$
\begin{array}{ccccccc}
K_{A_2}(BG \times \mathbf{S}^1) & \longrightarrow & K_{A_2}(\mathbf{S}(BG)) & & b(\rho) & & \\
\cup & & \Big\downarrow = & & \cap & & \\
b(\bar\rho) & & K_{A_2}^1(BG) & \longrightarrow & K_\varphi(BG) & \longrightarrow & K_{A_1}(BG)
\end{array}
$$

For an illustration let

$$\varphi : B(H) \longrightarrow B(H)/\mathrm{Comp}(H) = \mathcal{C}(\mathcal{H}) \tag{6.33}$$

be the natural projection to the Calkin algebra. As the bundles ξ_1 and ξ_2 are trivial the isomorphism $b(F)$ generates a continuous family of Fredholm operators, that is, an element of the group $K(BG)$:

$$K_\varphi(BG) \sim K(BG).$$

Thus we have a homomorphism

$$b : R(G; \varphi) \longrightarrow K(BG). \tag{6.34}$$

Relative representations for the projection (6.33) are called *Fredholm represen-tations* of the group G. Since the group $K(BG)$ is nontrivial, we might expect that the homomorphism (6.34) is nontrivial.

The classical finite dimensional representations can be considered as a special cases of Fredholm representations. In fact, let us substitute for the condition (6.30) the stronger one :

$$\rho_1 F = F \rho_2; \quad g \in G.$$

Then the representations ρ_1 and ρ_2 induce two finite dimensional representations in **Ker** F and **Coker** F. It is clear that

$$b(\textbf{Ker } F) - b(\textbf{Coker } F) = b(\rho_1, F, \rho_2) \in K(BG).$$

For example, consider the discrete group $G = \mathbf{Z}^n$. We showed earlier that for any finite dimensional representation ρ the associated vector bundle $b(\rho)$ is trivial, that is, the homomorphism

$$b : R(\mathbf{Z}^n) \longrightarrow K(B\mathbf{Z}^n) \tag{6.35}$$

is trivial.

However, if we enlarge the group of virtual representations by adding Fredholm representations the homomorphism (6.35) becomes an epimorphism. Moreover, the following theorem is true.

Theorem 60 *Let G be a group such that BG can be represented by a smooth complete Riemannian manifold with nonpositive section curvature. Then the homomorphism*
$$b : R(G, \varphi) \longrightarrow K(BG)$$
is an epimorphism.

When $G = \mathbf{Z}^n$ we have a special case of Theorem 60 since $B\mathbf{Z}^n$ is the torus with trivial sectional curvature. (see for example Mishchenko [28], Solov'yov [46], Kasparov [24])

6.4.2 Asymptotic finite dimensional representations

Another example of a very geometric variation of representation theory that may lead to a nontrivial homomorphism (6.29) is given by the so called asymptotic representations . Here we follow the ideas of Connes, Gromov and Moscovici [16].

Assume that

$$\rho : G \longrightarrow U(N)$$

is a map that is not actually a representation but is an approximation defined by the conditions:

1. $\rho(g^{-1}) = (\rho(g))^{-1}$, $g \in G$,

2. $\|\rho(g_1 g_2) - \rho(g_1)\rho(g_2)\| \le \varepsilon$ for $g_1, g_2 \in K \subset G$.

If K is a finite subset including a set of generators of the group G and the number ε is sufficiently small we can say that ρ is an *almost representation* of the group G. Then in the following section 3.4 we construct candidates for the transition functions on BG,

$$\phi_{\alpha\beta} = \rho \circ \psi_{\alpha\beta}, \tag{6.36}$$

where $\psi_{\alpha\beta}$ are the transition functions of the universal covering of BG. The problem here is that the functions (6.36) do not satisfy the cocycle condition (1.4). To rescue this situation we should make a small deformation of the functions (6.36) to make them into a cocycle.

To avoid certain technical difficulties let us change the definition slightly. Let

$$\|\rho\|_K = \sup_{g_1, g_2 \in K} \|\rho(g_1)\rho(g_2) - \rho(g_1 g_2)\|.$$

Consider an increasing sequence of integers

$$N_k : N_k < N_{k+1}, \ N_k \longrightarrow \infty$$

and a sequence of maps

$$\rho = \{\rho_k : G \longrightarrow U(N_k)\}. \tag{6.37}$$

The norms of matrices for different N_k can be compared by using the natural inclusions

$$U(N_k) \longrightarrow U(N_{k+1}) : U(N_k) \ni X \mapsto X \oplus 1_{N_{k+1}-N_k} \in U(N_{k+1}).$$

Then the sequence (6.37) is called an *asymptotic representation* if the following conditions hold:

1. $\rho_k(g^{-1}) = (\rho_k(g))^{-1}$, $g \in G$,

2. $\lim\limits_{k \to \infty} \|\rho_k\|_K = 0$,

3. $\lim\limits_{k \to \infty} \|\rho_{k+1}(g) - \rho_k(g)\| = 0$, $g \in K$

It can be shown that an asymptotic representation is actually a representation in a special C^*-algebra.

6.5 CONCLUSION

The applications of vector bundles are not exhausted by the examples considered above. We summarize further outstanding examples of their use.

6.5.1 Fiberwise homotopy equivalence of vector bundles

Let $\mathbf{S}(\xi)$ be the bundle formed by all unit vectors in the total space of a vector bundle ξ over a base X. Two vector bundles ξ and η are said to be *fiberwise homotopy equivalent* if there are two fiberwise maps

$$f : \mathbf{S}(\xi) \longrightarrow \mathbf{S}(\eta); \quad g : \mathbf{S}(\eta) \longrightarrow \mathbf{S}(\xi),$$

such that the two compositions

$$gf : \mathbf{S}(\xi) \longrightarrow \mathbf{S}(\xi); \quad fg : \mathbf{S}(\eta) \longrightarrow \mathbf{S}(\eta)$$

are homotopic to the identity maps.

This definition is justified by the following theorem (Atiyah, [4]):

Theorem 61 *If*

$$f : X \longrightarrow Y$$

is a homotopy equivalence of closed compact manifolds, $\nu(X)$ and $\nu(Y)$ are normal bundles (of the same dimension) of X and Y, respectively, then the bundles $\nu(X)$ and $f^*(\nu(Y))$ are fiberwise homotopy equivalent.

In particular, Theorem 61 is important in the classification of smooth structures of a given homotopy type. The classes of fiberwise homotopy equivalent bundles form a group $J(X)$ which is a quotient of the group $K(X)$. The problem of describing the group $J(X)$ or the kernel of the projection $K(X) \longrightarrow J(X)$ can be solved by using the Adams cohomology operations Ψ^k. This description is reduced to the so called the Adams conjecture. Namely, let $f(k) \in \mathbf{Z}$ be an arbitrary positive function , where $W_f(X) \subset K(X)$ is the subgroup generated by elements of the form

$$k^{f(k)} \left(\Psi^k - 1 \right) y, \ y \in K(X).$$

If

$$W(X) = \bigcap_f W_f(X).$$

Theorem 62 (Quillen [39], Sullivan [51])

$$J(X) = K(X)/W(X).$$

In particular, the group $J(X)$ is finite.

6.5.2 Immersions of manifolds

Let

$$f : X \longrightarrow Y$$

be an immersion of a compact manifold. Then, clearly, the normal bundle $\nu(X \longrightarrow Y)$ has dimension

$$\dim \nu(X \longrightarrow Y) = \dim Y - \dim X.$$

Hence, by the dimension restriction, some characteristic classes must be trivial. This observation leads to some criteria showing when one manifold cannot be immersed or embedded in another given manifold.

For example, for $n = 2^r$, the real projective space $\mathbf{RP}^n$ cannot be immersed into $\mathbf{RP}^{2n-2}$ since then the normal bundle $\nu = \nu(\mathbf{RP}^n \longrightarrow \mathbf{RP}^{2n-2})$ would have a nontrivial Stieffel–Whitney class $w_{n-1}(\nu)$. Hence, the dimension cannot equal $n - 2$.

More generally, let $\nu(X)$ be a normal bundle for which the dimension is determined by the dimension of Euclidean space $\mathbf{R}^N$ in which the manifold X is embedded. Put

$$k = \min \dim \nu_0(X)$$

where the minimum is taken over all possible decompositions

$$\nu = \nu_0(X) \oplus \bar{n},$$

where $\bar{n}$ is a trivial bundle. The number k is called the *geometric dimension* of the normal bundle $\nu(X)$. It is clear that the manifold X cannot be embedded in the Euclidean space $\mathbf{R}^{\dim X + k - 1}$.

6.5.3   Poincare duality in K–theory

To define Poincare duality we need to understand which homology theory is dual to K–theory. The trivial definition for the homology groups $K_*(X)$,

$$K_*(X) \overset{def}{=} \mathbf{Hom}\ (K^*(X), \mathbf{Z}),$$

is incorrect since we do not have exactness of the homology pair sequence.

We need a more delicate definition. Consider a natural transformation of functors

$$\alpha : K^*(X \times Y) \longrightarrow K^*(Y) \qquad\qquad (6.38)$$

which is a homomorphism of modules over the ring $K^*(Y)$.

Clearly, all such transformations form a $\mathbf{Z}_2$–graded group, which we denote by

$$K_*(X) = K_0(X) \oplus K_1(X). \qquad\qquad (6.39)$$

It can be shown that the groups (6.39) form a homology theory. When $Y = \mathrm{pt}$, the homomorphism (6.38) can be understood as the value of a homology element on the cohomology element:

$$\langle \alpha, y \rangle = \alpha(y).$$

Theorem 63 *Let X be an almost complex manifold. Then there is a homomorphism (Poincare duality)*

$$D : K_c^*(X) \longrightarrow K_*(X)$$

satisfying the condition

$$\langle D(x), y \rangle = \langle xy, D(1) \rangle$$

Atiyah ([5]) noticed that each elliptic operator $\sigma(D)$ defines an element of the homology K–theory dual to the element $[\alpha] \in K_c(T^*X)$. The construction is as follows. A continuous family of elliptic operators

$$\sigma(D) \otimes \eta : \Gamma(\xi_1 \otimes \eta_y) \longrightarrow \Gamma(\xi_2 \otimes \eta_y), \quad y \in Y$$

is constructed for each vector bundle η over $X \times Y$. Then the index of this family induces a homomorphism

$$\text{index} : K(X \times Y) \longrightarrow K(Y)$$

which is by definition an element of the homology group $K_*(X)$.

The notion of an elliptic operator can be extended as follows. Consider a Fredholm operator $F : H \longrightarrow H$. Assume that the algebra $C(X)$ acts on the Hilbert space H in such a way such that F commutes with this action, modulo compact operators. Let ρ denote the action of the algebra. Clearly, the pair (F, ρ) gives a well defined family of Fredholm operators

$$F_y = F \otimes \eta : H \otimes_{C(X)} \Gamma(\eta_y) \longrightarrow H \otimes_{C(X)} \Gamma(\eta_y), y \in Y,$$

and hence defines a homomorphism

$$\text{index} : K(X \times Y) \longrightarrow K(Y),$$

that is, an element

$$[F, \rho] \in K_*(X).$$

Moreover, if $f : X_1 \longrightarrow X_2$ is a continuous map then the pair $(F, \rho \circ f^*)$ defines an element of $K_*(X_2)$ for which

$$f_*([F, \rho]) = [F, \rho \circ f^*].$$

When $Y_2 = \text{pt}$,

$$K_*(X_2) = \mathbf{Z}$$

and

$$[F, \rho \circ f^*] = \text{index } F.$$

This last formula in some sense explains why, in the Atiyah–Singer formula, the direct image appears.

6.5.4   Historical notes

The study of bundles was probably started by Poincaré with the study of non-trivial coverings. Fibrations appeared in connection with the study of smooth manifolds , then characteristic classes were defined and for a long time they were the main tool of investigation. The Stieffel–Whitney characteristic classes were introduced by Stieffel [50] and Whitney [52] in 1935 for tangent bundles of smooth manifolds. Whitney [53] also considered arbitrary sphere bundles. These were mod 2 characteristic classes.

The integer characteristic classes were constructed by Pontryagin [36]. He also proved the classification theorem for general bundles with the structure group $O(n)$ and $SO(n)$ using a universal bundle over a universal base such as a Grassmannian.

For complex bundles, characteristic classes were constructed by Chern [15].

This was the beginning of the general study of bundles and the period was well described in the book of Steenrod [48]

A crucial point of time came with the discovery by Leray of spectral sequences which were applied to the calculation of the homology groups of bundles. Another was at the discovery by Bott [13] of the periodicity of the homotopy groups of the unitary and orthogonal groups.

From this point, vector bundles occupied an important place in the theory of bundles as it was through them that nontrivial cohomology theories were constructed. It was through Bott periodicity that the Grothendieck constructions became so significant in their influence on the development of algebraic K–theory.

This period was very rich: the problem of vector fields on the spheres was solved, J-functor was calculated, a homotopy description of the group representations was achieved. Among the mathematicians important for their contributions in this period are: Adams, Atiyah, Bott, Hirzebruch, Borel, Godement, Milnor and Novikov.

The most brilliant example of an application of K-theory was the Atiyah–Singer formula for the index of elliptic operator. This theory generated new wave of work and applications of K-theory.

The contemporary period may be characterized as a time in which K-theory became a suitable language for theoretical physics and related mathematical branches - representation theory, operator and Banach algebras, quantizations, and so on. The name of this new circle of ideas is "non commutative geometry". But it is not possible here to give an exposition that would do it justice .

INDEX

REFERENCES

[1] Adams J.F., *Vector fields on spheres.* — Ann.Math., Vol.**75**, (1962) p. 603–632.

[2] Adams J.F., *On the groups $K(X)$, I.* — Topology, Vol.**2**, (1963) p. 181–195. *II.* — Topology, Vol.**3**, (1965) p. 137–171. *III.* — Topology, Vol.**3**, (1965) p. 193–222.

[3] Anderson D.W., *The real K-theory of classifying spaces.* — Proc. Nat. Acad. Sci. U.S.A., Vol.**51**, (1964) p. 634–636.

[4] Atiyah M.F., *Thom complexes.* — Proc. London Math. Soc., Vol.**11**, (1961) No.3, p. 291–310.

[5] Atiyah M.F., *Global theory of elliptic operators.* — Proc. Intern. Conf. on Functional Analysis and Related Topics (Tokyo 1969). Tokyo, Univ. Tokyo Press, (1970) p. 21–30.

[6] Atiyah M.F., *K-theory.* — Benjamin, New York, (1967)

[7] Atiyah M.F., *Vector bundles and the Kunneth formula.* — Topology, Vol.**1**, (1963) p. 245–248.

[8] Atiyah M.F., Bott R., *On the periodicity theorem for complex vector bundles.* — Acta Math., Vol.**112**, (1964) p. 229–247.

[9] Atiyah M.F., Singer I.M., *The index of elliptic operators on compact manifolds.* — Bull. Amer. Math. Soc., Vol.**69**, (1963) p. 422–433.

[10] Atiyah M.F., Singer I.M., *The index of elliptic operators, I.* — Ann. of Math., Vol.**87**, (1968) No.2, p. 484–530. , *II.* — p. 531–545. , *III.* — p. 546–604.

[11] Atiyah M.F., Hirzebruch F., *Vector bundles and homogeneous spaces in differential geometry.* — Amer. Math. Soc. Proc. Symp. Pure Math., Vol.**3**, (1961) p. 7–38.

[12] Borel A., Hirzebruch F., *Characteristic classes and homogeneous spaces, I..* — Amer. J. Math., Vol.**80**, (1958) p. 458–538. , *II..* — Vol.**81**, (1959) p. 315–382. , *III..* — Vol.**82**, (1960) p. 491–504.

[13] Bott R., *The stable homotopy of the classical groups.* —

Ann. Math., Vol.**70**, (1959) p. 313–337.

[14] Bott R., *Lectures on K(X).* — Cambridge, Mass. Harvard Univ., (1962)

[15] Chern S.S., *On the multiplication in the characteristic ring of a sphere bundle.* — Ann. Math., Vol.**49**, (1948) p. 362–372.

[16] Connes A., Gromov M., Moscovici H., *Conjecture de Novicov et fibres presque plats.* — C.R.Acad.Sci. Paris, Vol.**310**, (1990) p. 273–277.

[17] Connes A., Moscovici H., *Cyclic Cohomology, The Novikov Conjecture and Hyperbolic Groups.* — Topology, Vol.**29**, (1990) No.3, p. 345–388.

[18] Godement R., *Topologie algébrique et theorie des faisceaux.* — Hermann & Cie, Paris, (1958)

[19] Hirzebruch F., *A Riemann-Roch theorem for differential manifolds.* — Seminar Bourbaki, (1959) p. 177.

[20] Hirzebruch F., *Topological Methods in Algebraic Geometry.* — Berlin, Springer-Verlag, (1966)

[21] Hirsch M.W., *Immersions of manifolds.* — Trans. Amer. Math. Soc., Vol.**93**, (1959) p. 242–276.

[22] Husemoller D., *Fiber Bundles.* — McGraw-Hill, (1966)

[23] Karoubi M., *K-Theory, An Introduction.* — Springer-Verlag, Berlin, Heidelberg, New York, (1978)

[24] Kasparov G., *Equivariant KK-theory and the Novikov conjecture .* — Invent. Math., Vol.**91**, (1988) p. 147–201.

[25] Kuiper N.H., *The homotopy type of the unitary group of Hibert space.* — Topology, Vol.**3**, (1965) p. 19–30.

[26] Milnor J., *Construction of universal bundles,I.* — Ann. Math., Vol.**63**, (1965) p. 272–284. ,*II.* — p. 430–436.

[27] Milnor J., Stasheff J.D., *Characteristic Classes.* — Annals of Math. Studies No. 76, Princeton , (1974)

[28] Mishchenko A.S., *Infinite dimensional representations of discrete groups and homotopy invariants of non simply connected manifolds.* — Soviet math. survey, Vol.**28**, (1973) No.2, p. 239–240.

[29] Mishchenko A.S., *On Fredholm representations of discrete groups.* — Funk. an. and its applications, Vol.**9**, (1975) No.2, p. 229–256.

[30] Mishchenko A.S., Fomenko A.T., *Index of elliptic operators over* C^*-*algebras* — Math. USSR – Izvestia, Vol.**15**, (1980) p. 87–112.

[31] Mishchenko A.S., Solov'yov Ju.P., *Representations of Banach algebras and Hirzebruch type formulas.* — Math.USSR– sbornik, Vol.**39**, (1981) p. 189–205.

[32] Novikov S.P., *Pontrjagin classes, the fundamental group, and some problems of stable algebra.* — In "Essay on Topology and Related Topics. Mémoires dédiés à Georges de Rham", Springer, (1970) p. 147–155.

[33] Novikov S.P., *Algebraic construction and properties of Hermitian analogues of* K-*theory over rings with involution from the point of view of Hamilton formalism,I. .* — Izvestia Acad. Nauk SSSR, sert. mat., Vol.**34**, (1970) p. 253–288. ,*II.* . — p. 475–500.

[34] Palais R., *Seminar on the Atiyah–Singer index theorem .* — Ann. of Math. Stud. 57, Princeton Univ. Press, Princeton, N.J., (1965)

[35] Paschke W., *Inner product modules over* B^*-*algebras.* — Trans. Amer. Math. Soc., Vol.**182**, (1973) p. 443–468.

[36] Pontryagin L.S., *Characteristic cycles of differentiable manifolds.* — Doklady Acad.Sci. USSR, Vol.**19**, (1938) p. 361–363.

[37] Pontryagin L.S., *Classification of some skew products.* — Doklady Acad. Sci. USSR, Vol.**43**, (1944) p. 91–94.

[38] Puppe D., *Homotopiemengen und ihre induzierten Abbildungen.* — Math. Z., Vol.**69**, (1958) p. 299–344.

[39] Quillen D., *Cohomology of groups.* — Actes Congres internat. Math., Vol.**2**, (1970) p. 47–51.

[40] Quillen D., *The Adams conjecture.* — Topology, Vol.**10**, (1971) p. 67–80.

[41] Rieffel, M.A. *Morita equivalence for* C^*-*algebras and* W^*-*algebras.* — J.Pure Appl. Algebra, Vol.**5**, (1974) p. 51–96.

[42] Rochlin V.A., *Three dimensional manifold is a boundary of four dimensional one.* — Dokl. Akad. Nauk SSSR, Vol.**81**, (1951) p. 355–357.

[43] Rochlin V.A., *New results in the theory of four dimensional manifolds.* — Doklady Akad. Nauk. USSR, Vol.**84**, (1952) p. 221–224.

[44] Serre J.-P., *Homologie singulière des espaces fibrès.* —

Ann. Math., Vol.**54**, (1951) p. 425–505.

[45] Shih Weishu, *Fiber cobordism and the index of elliptic differential operators.* — Bull. Amer. Math. Soc., Vol.**72**, (1966) No.6, p. 984–991.

[46] Solov'yov Ju.P., *Discrete groups, Bruhat–Tits complexes and homotopy invariance of higher signatures.* — Uspechi Math. Nauk, Vol.**31**, No.1, (1976) p. 261–262.

[47] Spanier E.H., *Algebraic Topology.* — N.Y., San Francisko, St. Luis, Toronto, London, Sydney, Mc Graw–Hill, (1966)

[48] Steenrod N.F., *The topology of fiber bundles.* — Princeton Univ. Press, Princeton, New Jersey, (1951)

[49] Steenrod N.F., *Classification of sphere bundles.* — Ann. Math., Vol.**45**, (1944) p. 294–311.

[50] Stiefel E., *Richtungsfelder und Fernparallelismus in Mannigfaltigkeiten.* — Comm. Math. Helv., Vol.**8**, (1936) p. 3–51.

[51] Sullivan D., *Geometric Topology.* — MIT, (1970)

[52] Whitney H., *Sphere spaces.* — Proc. Nat. Acad. Sci. U.S.A., Vol.**21**, (1935) p. 462–468.

[53] Whitney H., *On the theory of sphere bundles.* — Proc. Nat. Acad. sci. U.S.A., Vol.**26**, (1940) No.26, p. 148–153.

[54] Wood R., *Banach algebras and Bott periodicity.* — Topology, Vol.**4**, (1966) p. 371–389.

Other *Mathematics and Its Applications* titles of interest:

B.A. Plamenevskii: *Algebras of Pseudodifferential Operators.* 1989, 304 pp.
ISBN 0-7923-0231-1

Ya.I. Belopolskaya and Yu.L. Dalecky: *Stochastic Equations and Differential Geometry.* 1990, 288 pp. ISBN 90-277-2807-0

V. Goldshtein and Yu. Reshetnyak: *Quasiconformal Mappings and Sobolev Spaces.* 1990, 392 pp. ISBN 0-7923-0543-4

A.T. Fomenko: *Variational Principles in Topology. Multidimensional Minimal Surface Theory.* 1990, 388 pp. ISBN 0-7923-0230-3

S.P. Novikov and A.T. Fomenko: *Basic Elements of Differential Geometry and Topology.* 1990, 500 pp. ISBN 0-7923-1009-8

B.N. Apanasov: *The Geometry of Discrete Groups in Space and Uniformization Problems.* 1991, 500 pp. ISBN 0-7923-0216-8

C. Bartocci, U. Bruzzo and D. Hernandez-Ruiperez: *The Geometry of Super-manifolds.* 1991, 242 pp. ISBN 0-7923-1440-9

N.J. Vilenkin and A.U. Klimyk: *Representation of Lie Groups and Special Functions. Volume 1: Simplest Lie Groups, Special Functions, and Integral Transforms.* 1991, 608 pp. ISBN 0-7923-1466-2

A.V. Arkhangelskii: *Topological Function Spaces.* 1992, 206 pp.
ISBN 0-7923-1531-6

Kichoon Yang: *Exterior Differential Systems and Equivalence Problems.* 1992, 196 pp. ISBN 0-7923-1593-6

M.A. Akivis and A.M. Shelekhov: *Geometry and Algebra of Multidimensional Three-Webs.* 1992, 358 pp. ISBN 0-7923-1684-3

A. Tempelman: *Ergodic Theorems for Group Actions.* 1992, 400 pp.
ISBN 0-7923-1717-3

N.Ja. Vilenkin and A.U. Klimyk: *Representation of Lie Groups and Special Functions, Volume 3. Classical and Quantum Groups and Special Functions.* 1992, 630 pp. ISBN 0-7923-1493-X

N.Ja. Vilenkin and A.U. Klimyk: *Representation of Lie Groups and Special Functions, Volume 2. Class I Representations, Special Functions, and Integral Transforms.* 1993, 612 pp. ISBN 0-7923-1492-1

I.A. Faradzev, A.A. Ivanov, M.M. Klin and A.J. Woldar: *Investigations in Algebraic Theory of Combinatorial Objects.* 1993, 516 pp. ISBN 0-7923-1927-3

M. Puta: *Hamiltonian Mechanical Systems and Geometric Quantization.* 1993, 286 pp. ISBN 0-7923-2306-8

V.V. Trofimov: *Introduction to Geometry of Manifolds with Symmetry.* 1994, 326 pp. ISBN 0-7923-2561-3

Other *Mathematics and Its Applications* titles of interest:

J.-F. Pommaret: *Partial Differential Equations and Group Theory. New Perspectives for Applications.* 1994, 473 pp. ISBN 0-7923-2966-X

Kichoon Yang: *Complete Minimal Surfaces of Finite Total Curvature.* 1994, 157 pp. ISBN 0-7923-3012-9

N.N. Tarkhanov: *Complexes of Differential Operators.* 1995, 414 pp. ISBN 0-7923-3706-9

L. Tamássy and J. Szenthe (eds.): *New Developments in Differential Geometry.* 1996, 444 pp. ISBN 0-7923-3822-7

W.C. Holland (ed.): *Ordered Groups and Infinite Permutation Groups.* 1996, 255 pp. ISBN 0-7923-3853-7

K.L. Duggal and A. Bejancu: *Lightlike Submanifolds of Semi-Riemannian Manifolds and Applications.* 1996, 308 pp. ISBN 0-7923-3957-6

D.N. Kupeli: *Singular Semi-Riemannian Geometry.* 1996, 187 pp. ISBN 0-7923-3996-7

L.N. Shevrin and A.J. Ovsyannikov: *Semigroups and Their Subsemigroup Lattices.* 1996, 390 pp. ISBN 0-7923-4221-6

C.T.J. Dodson and P.E. Parker: *A User's Guide to Algebraic Topology.* 1997, 418 pp. ISBN 0-7923-4292-5

B. Rosenfeld: *Geometry of Lie Groups.* 1997, 412 pp. ISBN 0-7923-4390-5

A. Banyaga: *The Structure of Classical Diffeomorphism Groups.* 1997, 208 pp. ISBN 0-7923-4475-8

A.N. Sharkovsky, S.F. Kolyada, A.G. Sivak and V.V. Fedorenko: *Dynamics of One-Dimensional Maps.* 1997, 272 pp. ISBN 0-7923-4532-0

A.T. Fomenko and S.V. Matveev: *Algorithmic and Computer Methods for Three-Manifolds.* 1997, 346 pp. ISBN 0-7923-4770-6

A. Mallios: *Geometry of Vector Sheaves.* An Axiomatic Approach to Differential Geometry.
Volume I: Vector Sheaves. General Theory. 1998, 462 pp. ISBN 0-7923-5004-9
Volume II: Geometry. Examples and Applications. 1998, 462 pp.
ISBN 0-7923-5005-7
(Set of 2 volumes: 0-7923-5006-5)

G. Luke and A.S. Mishchenko: *Vector Bundles and Their Applications.* 1998, 262 pp. ISBN 0-7923-5154-1